The Amateur Plankton Researcher's Practical Guide

Albert Calbet

The Amateur Plankton Researcher's Practical Guide

How to Study Plankton at Home

 Springer

Albert Calbet 🅾
Institute of Marine Sciences - CSIC
Barcelona, Spain

ISBN 978-3-031-80247-8   ISBN 978-3-031-80248-5   (eBook)
https://doi.org/10.1007/978-3-031-80248-5

This Springer imprint is published by the registered company Springer Nature Switzerland AG
The registered company address is: Gewerbestrasse 11, 6330 Cham, Switzerland

If disposing of this product, please recycle the paper.

Disclaimer of Responsibility

The information and activities provided in this book are intended for educational purposes to foster interest in plankton sampling and experimentation by amateur scientists. While every effort has been made to ensure that the methods outlined are safe and accurate, the author and publisher are not responsible for any accidents, injuries, or damages that may arise from following these instructions.

Sampling in or near bodies of water can carry inherent risks. **Adult supervision is strongly advised at all times**, especially for minors and inexperienced individuals. Always follow safety precautions and local regulations, and respect the environment when collecting samples or conducting experiments.

By using this book, you acknowledge that you assume full responsibility for your safety and the safety of others participating in any related activities. When in doubt, seek advice or guidance from qualified professionals.

About the Guide

This practical guide is designed to help you study plankton at home, whether you are a student, educator, hobbyist, or citizen scientist. It will guide you through the entire process—from collecting samples in local water bodies to observing and documenting your findings, and even conducting basic experiments. While the guide introduces some concepts of plankton classification, and you can find representative images of the major plankton groups, it is not intended as a taxonomic reference.

Here is what you can expect:

- **A Brief Introduction to Plankton:** This chapter summarizes the essential knowledge, including the principal groups of plankton, their role in ecosystems, and the threats they face because of global changes. You will also gain insights into their seasonal variations and the environmental factors affecting their populations.
- **Getting Started:** Here, you will learn the essential equipment and materials needed to safely and effectively collect and study plankton.
- **Collecting Plankton Samples:** In this chapter, you will discover methods and tools for gathering and preserving samples for observation.
- **Observing Plankton:** Find out how to prepare samples, use a microscope, and identify common plankton groups.

- **Conducting Experiments:** Would you like to feel like a true scientist? This section will show you how to design and conduct simple experiments to explore plankton behavior and ecology, and how to analyze and interpret your data.
- **The Role of Molecular Techniques in Plankton Identification:** If you want to know about the latest technologies to study plankton, this chapter delves into more sophisticated methods, like DNA barcoding, environmental DNA, etc., for more detailed plankton studies.
- **Resources and Further Reading:** Here, you will access a list of books, online databases, and communities to deepen your understanding and connect with other plankton enthusiasts.
- **Glossary:** Use the glossary to familiarize yourself with key terms and refer to the appendices for useful tools like data sheets and identification images.
- **Practical Plankton Image Guide:** To complete the guide, you can find in this last section a library of pictures of the most abundant planktonic groups. The guide will be very useful to compare your organisms with a list of reference images. Also here, you will learn the anatomical and physiological characteristics of plankton that make each group unique.

By the end of this guide, I hope you will have the knowledge and skills to explore the captivating world of plankton and contribute to the understanding of these vital organisms.

Contents

About the Author

Albert Calbet My passion for plankton began when I was just ten years old. It all changed for me on December 31st when I received a simple plastic microscope as a birthday present from my parents. I can still vividly recall observing my first protists—it remains one of my most enduring memories. The sample came from a flower jar on our dining table, where I noticed tiny dots moving around on the water's surface. Placing a drop of that water on a glass slide and looking through the microscope, I was astonished by the new world that unfolded before me. Creatures like *Euglena viridis* (a flagellated microalgae) and *Stylonychia mytilus* (a flat ciliate) dominated the scene. From that moment on, I could not resist exploring every pond, vessel, river, or lake I came across. I used to carry small containers in my pockets (and still do) to collect water samples for later observation.

My interest extended beyond plankton to other forms of life, leading me to study Biology at the University of Barcelona, my hometown. There, I had the privilege of learning from inspiring educators who fueled my drive to pursue this career. Although working as a bartender, cook, and waiter on weekends and summers slowed down my studies, I eventually completed my degree and began my PhD at the Marine Sciences Institute (CSIC, Barcelona). I shifted my focus to marine zooplankton, primarily copepods, despite never having been particularly drawn to marine systems (probably due to my tendency to get seasick!). Yes, even marine biologists can struggle with seasickness. After four years of dedication, I successfully defended my PhD.

The next step in my journey was a postdoctoral position at the University of Hawaii, USA. I am not going to lie—I worked hard, but I also had an incredible experience. It was in Hawaii that I first became interested in marine microzooplankton, which has captivated me ever since, although I have not lost my enthusiasm for larger zooplankton.

Following my postdoc, I returned to Barcelona, spending a few years hopping from contract to contract before securing a permanent research position at the institution where I completed my PhD. Now, after more than three decades devoted to the study of plankton, I am passionate about sharing my knowledge with all kinds of audiences, especially the next generation of scientists. This is what motivated me to write this book, offering a practical and accessible guide to the fascinating world of plankton. I hope you enjoy it.

1

A Brief Introduction to Plankton

1.1 What Are Plankton?

Plankton (Fig. 1.1) are diverse collections of organisms that live in marine and freshwaters and are unable (most of the time) to swim against the current. They are a crucial part of aquatic ecosystems, serving as the foundation of the food web. Plankton are typically divided into several main categories:

- **Bacterioplankton** and **virioplankton**: Bacterioplankton play a crucial role in the ocean's health. They feed on organic matter, breaking it down and recycling important nutrients. This process supports the entire marine food web, helping to keep coral reefs, seagrass beds, and fish populations healthy. But there is more to the story. The ocean is also home to an even smaller group: viruses. These tiny entities, though often invisible to the naked eye, have a big impact on marine life. Viruses infect not only bacteria but also other tiny plankton. They help control planktonic populations and play a key role in recycling nutrients. When viruses infect bacterioplankton or other organisms, they can cause them to burst. This releases organic material back into

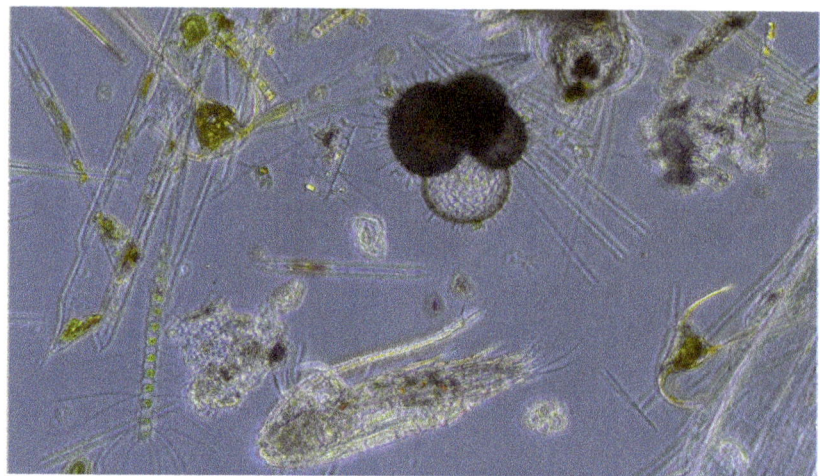

Fig. 1.1 Diverse plankton, including both phytoplankton and zooplankton

the water, which helps fuel new life and nutrient cycles. This process is essential for maintaining the balance of marine ecosystems.

- **Phytoplankton:** These are photosynthetic organisms, similar to terrestrial plants, and include groups like diatoms, cyanobacteria, and dinoflagellates, among others. They generate energy through photosynthesis, using sunlight to convert carbon dioxide (CO_2) and water into oxygen (O_2) and glucose. Phytoplankton are primary producers, forming the base of the aquatic food web. However, many of them are, at the same time, prey and predators. Akin to terrestrial carnivorous plants, we find a large array of mixotrophs within phytoplankton. This group will be referred to as mixoplankton.

- **Zooplankton:** These are heterotrophic organisms, meaning they feed on other plankton, including bacterioplankton, phytoplankton, other zooplankton, or detritus. Zooplankton include small crustaceans like copepods and cladocerans, jellyfish and other jelly organisms, fish larvae, and protozoans (unicellular), among others. They play a critical role in transferring energy up the food chain to larger predators, such as fish.

Plankton vary widely in size, from microscopic organisms that can only be seen under a microscope to larger species visible to the naked eye.

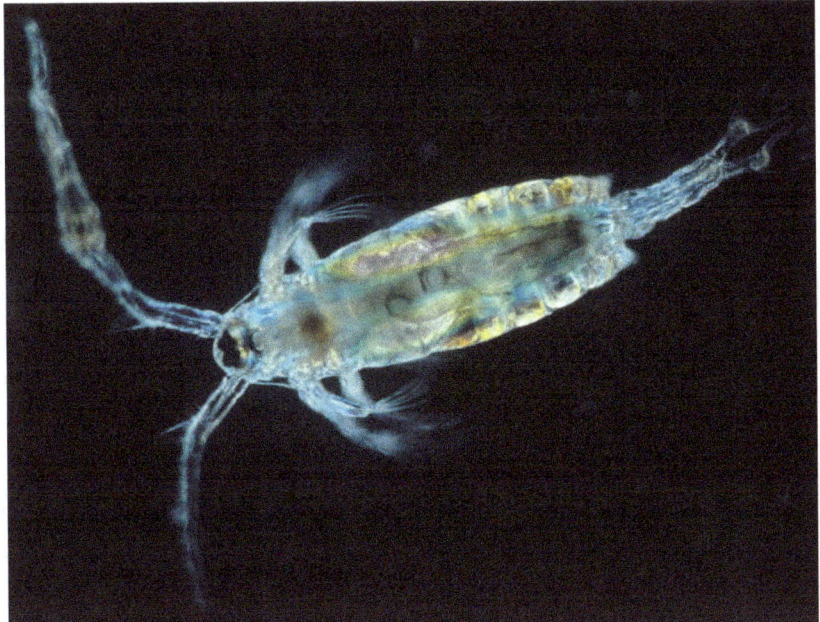

Fig. 1.2 Marine copepod. *Labidocera wollastoni*

Within a few drops of water (1 milliliter), you will find tens of millions of viruses, millions of bacteria, and hundreds of thousands of flagellates. There are also thousands of microalgae, some ciliates, dinoflagellates, and even, occasionally, one tiny crustacean such as copepods (Fig. 1.2) or cladocerans. It is a bustling microcosm, teeming with life in numbers that stretch the limits of imagination.

Scientists also classify them according to size. At the smallest end of the scale, you have the **femtoplankton**, which are so small you would need a powerful electron microscope to see them. They are like the dust particles of the ocean, including mostly viruses. Next in size, we have the **picoplankton**—so tiny that they are invisible to the naked eye. These microscopic powerhouses, like certain bacteria and tiny algae, are usually less than 2 micrometers[1] across. Following up are the **nanoplankton**, still

---

[1] A **micrometer** (μm) is a unit of length that is incredibly small—there are 1000 micrometers in a single millimeter, and about 25,400 micrometers in an inch.

incredibly small, but slightly larger than picoplankton. They range from 2 to 20 micrometers, which means you would need a microscope to see them, but they are crucial players in the ocean's food web, often consisting of small algae and protozoa. Moving up in size, we meet the **microplankton**. These are between 20 and 200 micrometers, about the width of a human hair. They are big enough that with a bit of help from a microscope, you can actually start to see their shape and behavior. Many of the classic plankton images, like diatoms and dinoflagellates, fall into this category. Then, we have the **mesoplankton**, which range from 200 micrometers to 20 millimeters—now we are getting to the point where some are visible without a microscope! These include small crustaceans like copepods and cladocerans, which are key food sources for many larger marine animals. Finally, there is the **macroplankton**, which can be up to 20 centimeters, and the **megaplankton**, anything larger than that. These are the giants of the plankton world, including some jellyfish and large tunicate colonies, such as salps. Despite their size, they are still considered plankton because they drift with the ocean currents. Therefore, from the tiny femtoplankton to the large megaplankton, plankton comes in all shapes and sizes, forming the very foundation of the ocean's ecosystem.

Finally, there is another way to classify plankton, particularly zooplankton, according to their life cycle: **Holoplankton** are organisms that remain in the planktonic state throughout their entire life cycle. They spend their whole lives drifting in the water column, whether in oceans, seas, or freshwater bodies. These organisms are integral to the aquatic food web and are adapted to thrive in this floating lifestyle. Major groups include copepods, rotifers, krill, etc.

Meroplankton, on the other hand, are temporary residents of the planktonic environment. They spend only a part of their life cycle as plankton before undergoing metamorphosis into different life stages, often transitioning to benthic (bottom-dwelling) or nektonic (active swimming) forms. Major groups are crustacean larvae, such as nauplii or zoea of crabs, shrimps and lobsters, fish larvae and echinoderm larvae (Fig. 1.3).

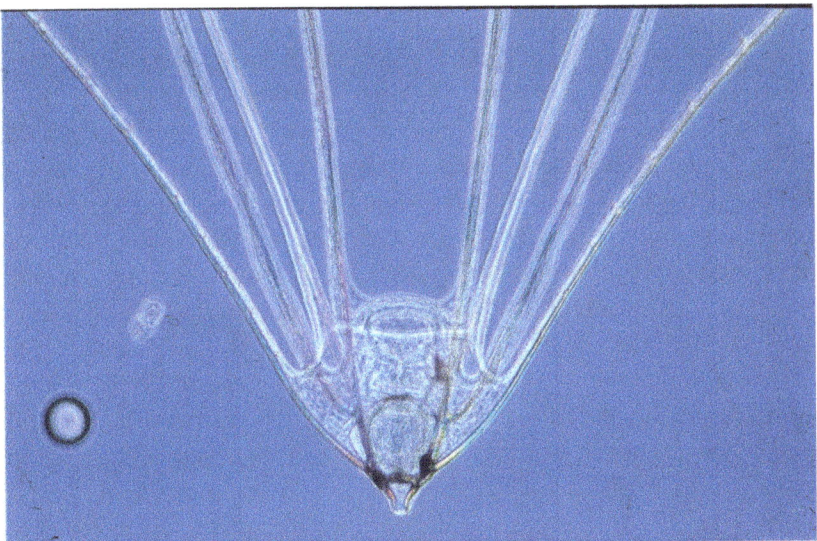

Fig. 1.3 Brine star larva

1.2 Plankton's Roles in the Food Web

The microscopic community within freshwaters and oceans plays a crucial role in sustaining life. Viruses regulate bacterial populations, preventing overgrowth, while bacteria decompose organic matter, fostering a nutrient cycle vital for the health of marine environments. Without the diligent work of bacteria, the ocean, lakes, and rivers would become graveyards of organic debris and dead bodies.

Phytoplankton, with their simple, plant-like structure, perform a task of monumental importance: they convert inorganic nutrients, by-products of bacterial activity, into living tissues through photosynthesis and other chemical reactions. In doing photosynthesis, phytoplankton bind CO_2 with water, channeling sunlight into the life-sustaining process and releasing oxygen as a beneficial by-product. Although it is commonly believed that marine algae contribute half of the oxygen we breathe, this is a nuanced truth. While they have certainly contributed to most atmospheric oxygen over the eons and conduct about half of the world's photosynthesis, the oxygen generated in aquatic environments is largely consumed there, with only a small portion diffusing into our atmosphere. However, they play a significant role in controlling CO_2 atmospheric levels and, therefore, in mitigating global warming.

From unicellular ciliates and dinoflagellates to the larvae of worms and mollusks, and to relatively larger organisms like certain crustaceans and tunicates, the consumers of phytoplankton are diverse. Among them, copepods stand out in marine ecosystems as tiny crustaceans with the largest population, outnumbering even the ubiquitous insects on land. Copepods are a primary food source for fish and, occasionally, for larger ocean dwellers like whales, although whales prefer the larger krill. Before becoming prey, copepods themselves feed on ciliates and algae. In freshwater ecosystems, the role of copepods is mostly performed by cladocerans, such as *Daphnia* (Fig. 1.4) species and similar genera. Copepods are present in these less salty waters but usually play more marginal roles.

The complex interactions and consumption patterns among these organisms create a web of life, where each member relies on others within the trophic system. This food web is more than just a hierarchy of who eats whom; it is also a cycle of nutrient recycling conducted within the microbial loop. This marine recycling is a prime example of how the concept of reuse and sustainability has been a natural part of Earth's processes for millions of years—a system where almost nothing is wasted.

The ocean's smallest creatures, therefore, hold a monumental position in our world, influencing everything from the air we breathe to the fish that grace our tables. They remind us that in the vast blueprint of life, even the tiniest entities have roles that are anything but small.

1.3   Seasonal and Spatial Variations in Plankton

Plankton populations vary seasonally and across different ecosystems. In most temperate aquatic systems, a seasonal pattern recurs annually with minimal variation.

In spring, as sunlight intensifies and water temperatures rise, plankton populations surge. Phytoplankton begin to bloom (Fig. 1.5), sometimes

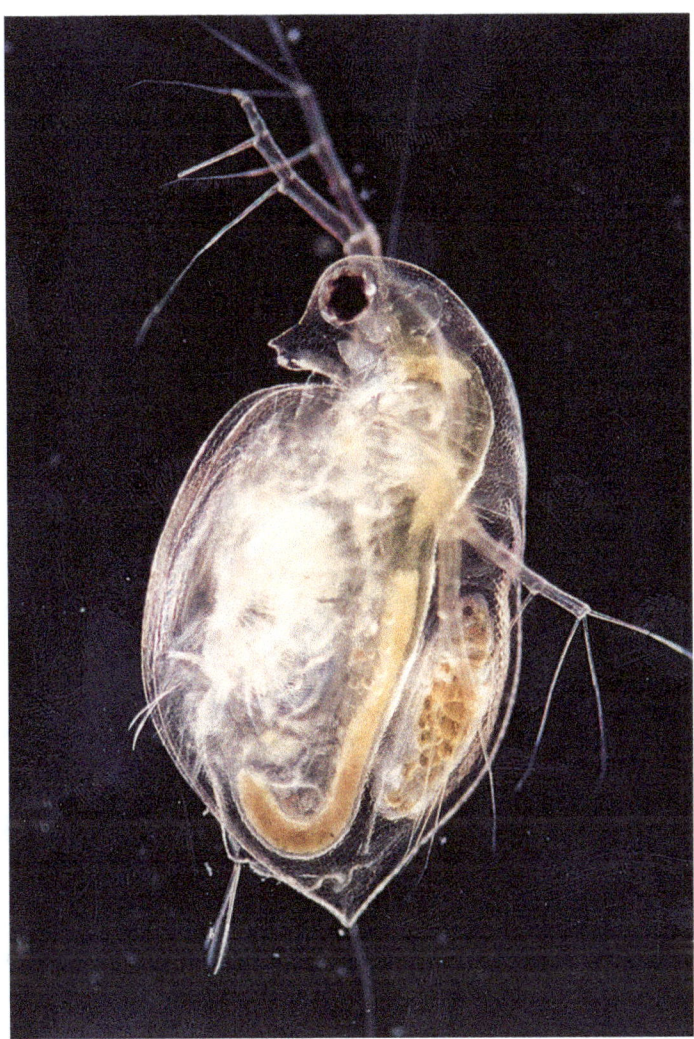

Fig. 1.4 *Daphnia pulex*

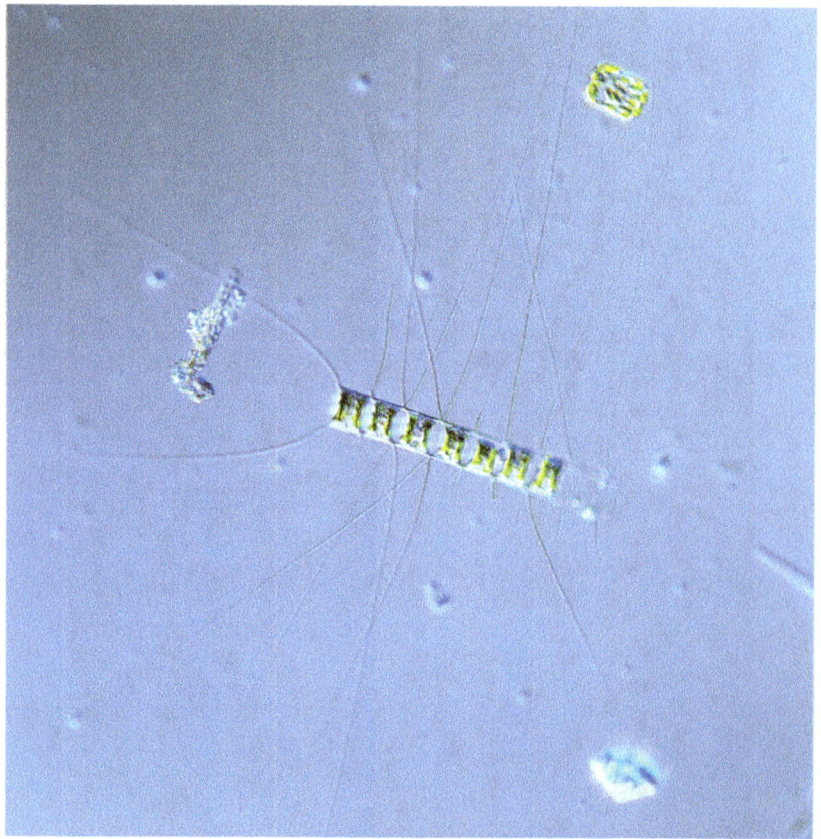

Fig. 1.5 Diatom chains are typical during spring blooms

coloring the water with vibrant greens. This spring bloom, observed in both marine and freshwater environments, provides an abundant food source for herbivorous zooplankton. These tiny creatures multiply rapidly, thriving in sudden abundance.

As summer arrives, the dynamic changes. In the ocean, lakes, and large ponds, nutrient levels near the surface start to decline, leading to a pause in phytoplankton growth. The summer sun heats the water, often causing it to stratify, creating layers of warm surface water over cooler deep water. Plankton adapt accordingly, with different species thriving in each layer, forming a vertical tapestry of life.

During autumn, a period of transition occurs. As temperatures drop, water mixing resumes. In the ocean, this mixing brings nutrients back to the surface, sparking a smaller but still significant phytoplankton bloom. In freshwater environments, cooler weather often results in similar mixing, redistributing nutrients and allowing different plankton species to emerge as others wane. It is a time of shifting roles and preparations for the impending winter.

Winter is a quieter season, with plankton populations dwindling in many parts of the ocean and freshwater bodies. The cold, darker months are a period of survival. Yet, even in this light-limited period, life persists—certain hardy species continue drifting through the cold waters, ready to ignite the next cycle of life when the seasons change again.

However, this seasonal cycle is only part of the story. Spatial variations—the different environments where plankton reside—add another layer of complexity. In the open ocean, also known as the pelagic zone, the waters are deep, vast, and often low in nutrients. This region, far from land, is sometimes called the "plankton desert" due to its relative scarcity of life compared to more nutrient-rich areas. Here, phytoplankton must adapt to the limited availability of nutrients, often relying on small species capable of surviving on minimal resources. Zooplankton, in turn, must efficiently graze on whatever prey they can find. Despite these challenges, the open ocean is home to some of the most resilient and specialized plankton communities on Earth, drifting with the currents across endless blue expanses.

Closer to shore, coastal waters present a stark contrast. These nutrient-rich regions, where land meets the sea, benefit from the influx of minerals from rivers, estuaries, and sometimes upwelling currents that bring deep ocean nutrients to the surface. Here, plankton communities (Fig. 1.6) explode in richness and abundance. Phytoplankton in coastal waters bloom rapidly, providing a food source for a wide array of zooplankton. This abundance, in turn, supports a thriving ecosystem, from small fish to large marine mammals. Coastal waters are like fertile gardens, with dense plankton populations that shift with the tides, seasons, and nutrient flow from the land.

In certain parts of the ocean, there are special regions known as upwelling zones, where deep, cold, nutrient-rich waters rise to the surface. These

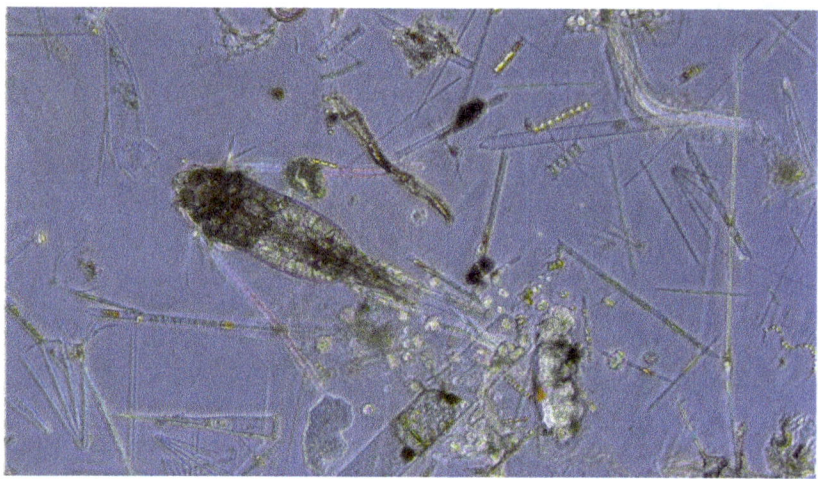

Fig. 1.6 Coastal plankton community. NW Mediterranean

areas, often found along coastlines where winds push surface waters away from the land, create plankton oases in the ocean. The sudden availability of nutrients sparks massive phytoplankton blooms, which in turn attract a diverse array of zooplankton. These oases are hotspots of life, drawing in fish, birds, and marine mammals that come to feast on the abundant food. Upwelling zones are crucial for global marine productivity and are among the most biologically active regions on the planet.

Estuaries and bays, where freshwater from rivers meets the salty ocean, are dynamic and productive environments. These areas are often rich in nutrients and provide a sheltered environment, making them ideal nurseries for many marine species. Plankton in estuaries must cope with varying salinity levels as the mix of fresh and saltwater shifts with the tides. Despite these challenges, estuaries are bustling with life, supporting diverse plankton communities that, in turn, feed young fish, shellfish, and other marine organisms that rely on these protected waters to grow before venturing into the open ocean.

Inland, freshwater bodies like lakes and rivers present entirely different challenges for plankton. These environments are often more isolated and smaller in scale than the ocean, leading to unique plankton communities (Fig. 1.7) that are highly adapted to their specific conditions. Lakes, for

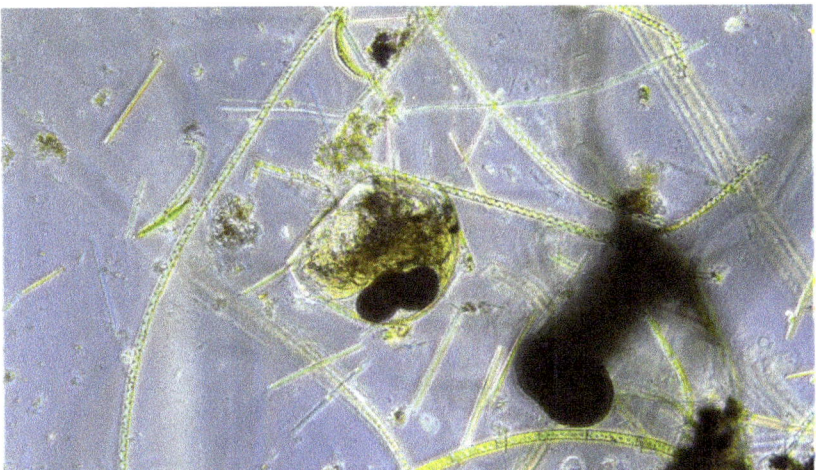

Fig. 1.7 Freshwater plankton community

example, can have stratified layers of water with different temperatures and nutrient levels, creating distinct habitats for different types of plankton. Rivers, with their constant flow, carry plankton downstream, often leading to a continuous renewal of species that must adapt to life in moving water. Each lake and river can have its own unique plankton population, shaped by factors like water chemistry, temperature, and flow rate.

Plankton communities also vary dramatically between polar and tropical regions. In the icy waters of the Arctic (Fig. 1.8) and Antarctic, plankton must survive in extremely cold conditions with long periods of darkness in winter and endless daylight in summer. Here, plankton blooms are often seasonal, peaking during the brief summer months when the sun returns. In contrast, tropical waters, which are warm and sunlit year-round, can have more consistent plankton populations, although these regions are often nutrient-poor and rely on upwelling or river input to boost productivity. The differences between polar and tropical plankton communities highlight the adaptability of these organisms to extreme environments.

In summary, the spatial scale of plankton variations reveals the incredible adaptability and diversity of these tiny organisms. From the vast, nutrient-poor open ocean to the nutrient-rich coastal waters, from the

Fig. 1.8 High Arctic in summer

dynamic estuaries to the isolated freshwater lakes, plankton communities are shaped by their environment in fascinating ways. Each location offers its challenges and resources, creating a complex mosaic of plankton life that is essential to the health of aquatic ecosystems around the world.

1.4   Where Are Plankton When They Are Not There?

As stated in the previous section, it is evident that the species composition in the water is not constant. In some seasons, a particular species might be abundant, while in others, it may be nearly absent or completely missing. This seasonal variation raises a simple yet intriguing question: where do plankton go when they are not present in the water column? Are they gone, or are they merely hidden from view? The truth is, while some species are present all year round, although under low abundances, many plankton species have developed remarkable survival strategies, allowing

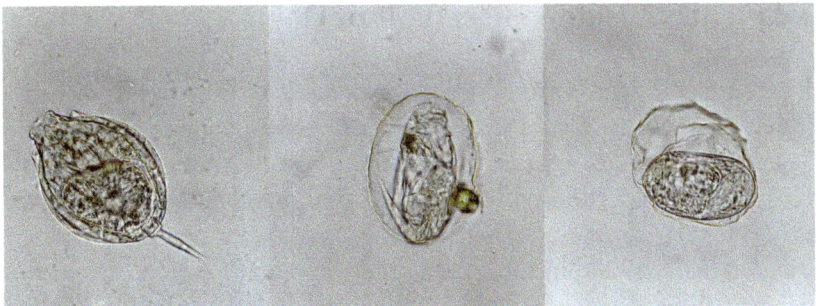

Fig. 1.9 Freshwater rotifer entering in its resting stage

them to persist even during unfavorable conditions, and a significant part of this story is found not in the water column, but in the sediments below.

The Role of Resting Stages

One very fascinating aspect of plankton biology is their ability to enter resting stages, specialized forms that allow them to endure periods of environmental stress. These resting stages are like time capsules—hardened, resistant forms such as cysts, spores, or eggs that can remain viable for months, years, or even decades. When conditions in the water column deteriorate, such as during the colder months of winter or during periods of low food availability, many plankton species either sink to the bottom and enter this dormant state (Fig. 1.9) or produce resting eggs. These resting stages are a crucial part of the life cycle for a wide variety of planktonic organisms, including both phytoplankton and zooplankton.

The concept of resting stages is not unique to plankton; many terrestrial plants also produce seeds that lie dormant in the soil until favorable conditions, such as sufficient rainfall and sunlight, trigger germination. Similarly, in aquatic environments, the resting stages of plankton wait in the sediments for the right conditions—such as warmer temperatures or increased nutrient availability—to hatch or resume their active forms. This ability to "pause" their life cycle enables plankton to survive through periods that would otherwise be lethal, ensuring the continuity of their populations.

Sediments: A Hidden Reservoir of Life

The sediments at the bottom of the ocean, lakes, or even ponds serve as a hidden reservoir for plankton. In these sediments, a remarkable number of planktonic species exist in their resting stages, forming a "seed bank" of sorts, much like the seed banks in soil that preserve the genetic diversity of terrestrial plants. These resting stages can remain viable for extended periods, and in some cases, they represent a significant portion of the total plankton population in a given area. When environmental conditions improve, these dormant forms hatch and rise back into the water column, contributing to the renewal of plankton populations.

However, it is important to note that the duration of dormancy can vary greatly depending on the species and environmental factors. Some plankton may remain dormant for just a few weeks, while others can persist for several years; the record for plankton is 600 years for a freshwater cladoceran. The timing of their emergence is carefully synchronized with favorable conditions, such as specific light levels, temperatures, or the availability of nutrients. This synchronization ensures that when the plankton return to the water column, they do so under conditions that maximize their chances of survival and reproductive success.

The Need for Sediment Sampling

Despite the importance of resting stages, traditional plankton research has often focused almost exclusively on the water column, largely neglecting the role of sediments. While studying plankton in the water column can provide valuable information about current populations, it offers only a snapshot of the overall dynamics. Without considering the sediments and the resting stages within them, researchers risk missing a crucial aspect of the plankton's life cycle.

Understanding plankton ecology requires a more holistic approach that includes sediment sampling. By examining the sediments, researchers can uncover the dormant stages of plankton and gain insight into the long-term dynamics of plankton populations. This approach can reveal species that may not be present in the water column during certain seasons, as well as provide clues about the conditions that trigger their re-emergence. In this sense, sediments offer a window into the past and future of plankton communities, allowing us to better understand how they respond to environmental changes.

1.5 Environmental Factors Affecting Plankton

The world of plankton is like an ever-changing landscape, where tiny drifters are constantly influenced by the environment around them. Imagine them as delicate dancers on a vast stage, responding to the shifting lights, temperatures, and currents that direct their movements. These environmental factors—light, temperature, nutrients, and currents—act as both the stage and the script, shaping the lives of plankton in ways that are as dynamic as they are crucial.

Light: The Power of the Sun

For many plankton, especially phytoplankton, light is everything. It is their energy source, the key to photosynthesis, and the driving force behind their existence. In the bright surface waters, where sunlight is abundant, phytoplankton thrive, converting sunlight into energy and forming the base of the marine food web. But light is not just about abundance; it is about quality too. As one dives deeper into the ocean, the light changes, filtering out certain wavelengths and leaving others. Different species of phytoplankton are adapted to these variations—some flourish in the bright blues of shallow waters, while others thrive in the dim greens of the depths. In polar regions, the sun's seasonal presence, with its months of darkness and endless daylight, creates dramatic pulses of plankton life, with explosive blooms in the spring and summer followed by quiet winters.

Light also drives the vertical migration of zooplankton (Fig. 1.10). As the sun sets, zooplankton begin to rise from the deep, dark waters where they have been hiding during the day. They swim upward, sometimes covering hundreds of meters, to feast under the cover of the night. This

Fig. 1.10 The copepod *Calanus* is a strong vertical migrator

upward movement is driven by their need to feed. Near the surface, they find an abundance of food, such as algae and protozoans. However, during daylight hours, this journey would be too risky, as the bright light makes them vulnerable to visual predators, like fish. Therefore, with the first rays of sunlight, zooplankton begin their descent back into the deep, retreating to safety before the day's predators can spot them. This daily rhythm of rising and sinking is not just a survival strategy—it is also a crucial part of the marine ecosystem. It influences nutrient cycles, the feeding patterns of larger animals, and even carbon storage in the ocean.

Temperature: The Ocean's Thermostat

Temperature is another key player, setting the rhythm of life for plankton. Warm waters, like those found in tropical regions, speed up metabolic rates, leading to faster growth and reproduction for many plankton

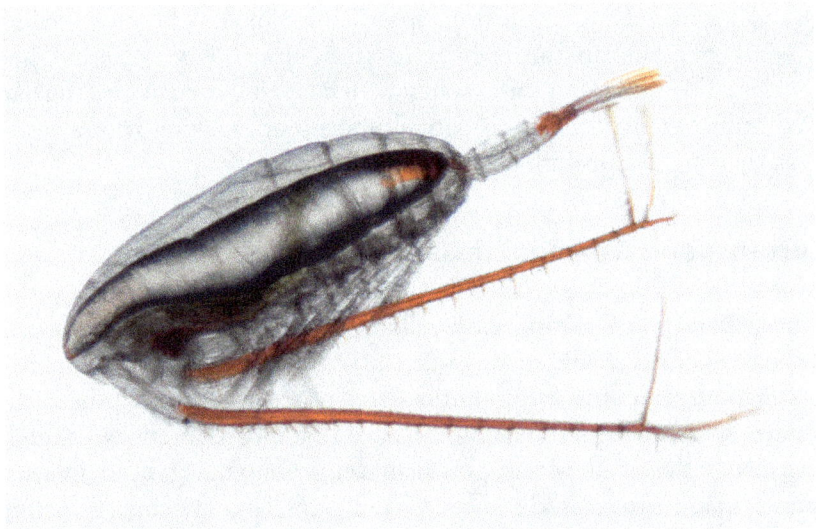

Fig. 1.11 *Calanus hyperboreus* is a species of copepod adapted to the cold conditions of the Arctic

species. However, warm waters can also be challenging—if they become too warm, stratification can occur, where the water layers do not mix well, trapping nutrients in the depths and leaving surface plankton hungry. In contrast, cooler waters, often found in temperate and polar regions, are rich in nutrients, thanks to the mixing of water layers. This mixing brings nutrients up from the deep, fueling massive plankton blooms, especially during the spring and fall. However, cold waters slow down metabolism, meaning that plankton in these regions (Fig. 1.11) often grow more slowly but can thrive for longer periods. Temperature is not just a backdrop—it is a thermostat that controls the pace and intensity of the plankton's life cycle.

Nutrients: The Fertilizer of the Sea

Nutrients are the food for plankton, the invisible sustenance that determines whether they bloom or wither away. Phytoplankton, in particular, respond quickly to a sudden influx of nutrients, creating massive blooms that can sometimes be seen from space. These blooms become feeding grounds for zooplankton, which in turn attract larger marine life. Nutrient availability can shift rapidly with changing currents, seasons, or even human activities such as agricultural runoff, making it one of the most dynamic factors influencing plankton life.

Currents and Mixing: The Ocean's Conveyor Belts

Currents and water mixing act like giant conveyor belts, moving plankton across the ocean and mixing nutrients throughout the water column. In some cases, currents bring nutrient-rich waters from the depths to the surface, sparking plankton blooms in areas that might otherwise be infertile. In other cases, they transport plankton from one region to another, spreading life across vast distances. Ocean gyres, large systems of rotating currents, can trap plankton in their slow-moving waters, creating regions of high productivity, while strong currents along coastlines can sweep plankton communities away, distributing them along new areas. Water mixing, driven by winds, tides, and temperature differences, ensures that nutrients are constantly cycled, creating opportunities for plankton to thrive even in seemingly unlikely areas. The constant motion of the ocean is a vital force, shaping where and how plankton live.

Human Impact: The Unseen Hand

In recent times, human activities have become another powerful force affecting plankton. Climate change is warming the oceans, altering current patterns, and changing the availability of nutrients, all of which impact plankton in ways that scientists are still striving to fully understand. Ocean acidification, driven by increased CO_2 levels, affects the ability of some plankton, particularly those with calcium carbonate shells, to survive and thrive. Pollution (Fig. 1.12), from plastics to chemical runoff, can harm plankton directly or disrupt the delicate balance of the ecosystems they inhabit. Even overfishing can have an indirect impact, as removing large predators can lead to changes in the plankton

Fig. 1.12 Coastal industries and oil refineries in Jurong Island, Singapore

community, disrupting the entire food web. Human influence is now a major factor in the lives of plankton, adding a new layer of complexity to their already intricate world.

Climate change is perhaps the most significant human impact on plankton today. As global temperatures rise, the oceans are warming too, setting off a cascade of effects that ripple through the marine food web. Warmer waters may seem like a minor inconvenience, but for plankton, they are a game-changer. The warming seas affect not only the temperature of the water but also the stratification of the ocean layers. Warmer surface waters can create a stronger barrier between the nutrient-rich deep waters and the sunlit surface waters where phytoplankton live. Without these nutrients, phytoplankton struggle to survive, leading to smaller blooms and less food for the zooplankton that depend on them.

The impacts of climate change do not stop there. Changing temperatures can also shift ocean currents, altering the distribution of plankton across the globe. Some species may find themselves in waters that are too warm, forcing them to move to cooler regions, if they can. Others may be unable to adapt, leading to declines in populations that have ripple effects throughout the food web. For example, in the Arctic, warming temperatures are leading to earlier ice melts, which disrupts the timing of plankton blooms. These blooms are crucial for the survival of many Arctic species, including fish, seabirds, and marine mammals. The delicate timing between plankton blooms and the life cycles of these species is altered, potentially leading to food shortages and population declines.

As humans continue to burn fossil fuels, the levels of CO_2 in the atmosphere are rising, and the oceans are absorbing much of this excess CO_2. This process leads to ocean acidification, a phenomenon where the increased CO_2 lowers the pH of seawater, making it more acidic. For many marine organisms, including plankton, this is a dangerous development.

Ocean acidification poses a particular threat to plankton with calcium carbonate shells, such as coccolithophores, foraminifera, and certain types of zooplankton like pteropods, often referred to as "sea butterflies." These tiny organisms rely on calcium carbonate to build and maintain their shells. As the water becomes more acidic, it reduces the availability of carbonate ions, which are essential for the formation of these shells. The result is thinner, weaker shells that make these plankton more vulnerable to predation and environmental stress.

The implications of ocean acidification extend far beyond individual species. Coccolithophores, for instance, are not only crucial in the marine food web but also play a significant role in the global carbon cycle. They sequester carbon dioxide by incorporating it into their shells, which eventually sink to the ocean floor, locking away carbon for centuries. As acidification hampers their ability to thrive, this natural carbon sequestration process is disrupted, potentially exacerbating climate change by leaving more CO_2 in the atmosphere.

Moreover, the decline of calcifying plankton can have cascading effects on the entire marine ecosystem. Many marine species, including commercially important fish and shellfish, depend on plankton at some stage of their life cycle. A reduction in plankton populations can lead to less food for these species, ultimately impacting fisheries and the human communities that rely on them.

Pollution, particularly from plastics and chemical runoff, is another human-induced stressor that is impacting plankton. Plastics, which have become ubiquitous in the world's oceans, break down into tiny particles known as microplastics. Although it is not yet clear if these microplastics are significantly ingested by plankton, the chemical substances they may leak, such as plasticizers and flame retardants, can have harmful effects. Chemical pollutants, such as pesticides, heavy metals, and industrial runoff, also pose a significant threat. These toxins can accumulate in the water, poisoning plankton and disrupting their reproductive processes. In some cases, the accumulation of these substances in plankton can scale up the food web, affecting top consumers like polar bears, large carnivorous fish, and even humans. Emerging pollutants, including pharmaceuticals, personal care products, and nanomaterials, represent a new wave of contaminants. These compounds, often unregulated, can enter the ocean through wastewater discharge and runoff. They may interfere with plankton's physiological functions, growth, and behavior, compounding the impact of traditional pollutants and adding another layer of risk to marine ecosystems.

Overfishing is another human activity that indirectly impacts plankton by disrupting the balance of marine ecosystems. Large fish and marine predators play a critical role in controlling the populations of zooplankton and other small marine organisms. When these predators are removed from the ecosystem, zooplankton populations can grow unchecked, leading to imbalances that can affect the entire food web, including phytoplankton populations, nutrient cycles, and the health of the marine environment. This trophic imbalance can result in cascading effects, where the unchecked growth of zooplankton may lead to overgrazing on phytoplankton. As phytoplankton are key producers in the ocean's food web and a major driver of global oxygen production and carbon sequestration, their depletion can weaken the base of the food chain, destabilizing ecosystems and reducing the productivity of fisheries. Additionally, overfishing alters nutrient dynamics, as the removal of predatory species can change the recycling of nutrients like nitrogen and phosphorus, which are vital for plankton growth. In some cases, the removal of large predators and fish species can create ecological niches that opportunistic species, such as jellyfish, are able to exploit. Without predators to keep their populations in check, jellyfish can proliferate, leading to large blooms that further disrupt the balance of the marine ecosystem. Jellyfish blooms can outcompete fish larvae and other zooplankton for food, worsening the decline of fish populations and intensifying the impact on the food web. The takeover by jellyfish is a clear example of how overfishing can trigger unexpected shifts in marine ecosystems, often favoring species that thrive in degraded environments, thereby compounding the ecological damage.

In conclusion, human activities are creating unprecedented changes in the world's oceans, and plankton, the tiny but mighty foundation of marine ecosystems, are feeling the impacts. Addressing these issues requires concerted global efforts, from reducing carbon emissions to protecting marine habitats and regulating pollutants. The health of plankton populations is not just a matter for marine biologists; it is a crucial concern for the health of our planet.

1.6   Importance of Studying Plankton

Plankton may be small, but their significance to our planet is enormous. Understanding these tiny organisms goes beyond peering into a drop of water under a microscope; it is about unlocking the secrets of ecosystems that sustain life on Earth. The study of plankton (Fig. 1.13) is essential for several compelling reasons, each revealing a different layer of their importance in the grand tapestry of our planet's ecology.

Ecological Significance

By studying plankton, scientists gain insights into the health and stability of ecosystems. Changes in plankton populations can signal shifts in water temperature, nutrient availability, and other environmental factors that affect biodiversity. For instance, a decline in phytoplankton might suggest a disruption in the availability of essential nutrients, potentially

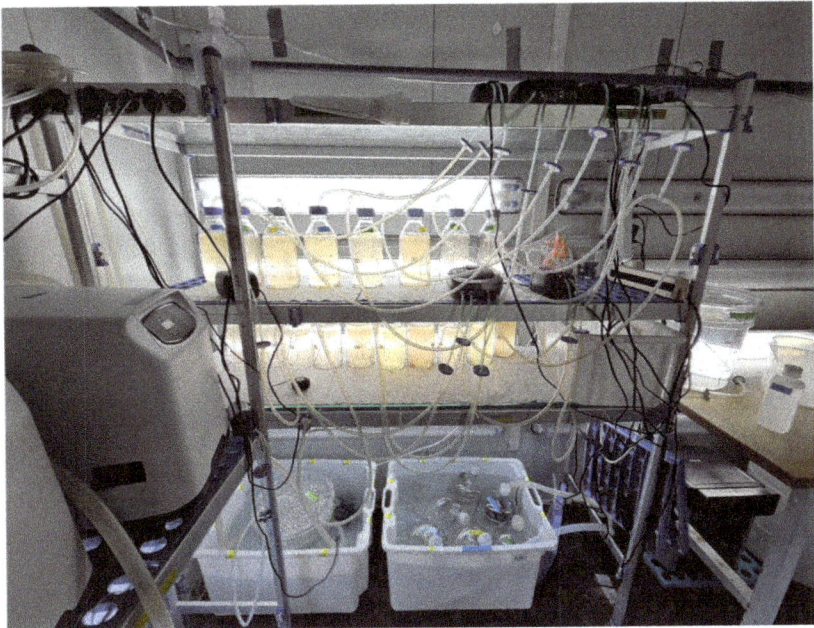

Fig. 1.13   Laboratory experiment involving plankton

leading to declines in fish populations and, ultimately, the species that depend on those fish. Conversely, an explosion of certain plankton species might indicate a nutrient overload, often resulting from pollution, which can lead to harmful algal blooms that devastate marine life.

Understanding plankton dynamics allows scientists to predict and manage the health of entire ecosystems. By monitoring plankton, researchers can identify early signs of ecosystem stress, helping to guide conservation efforts and protect the biodiversity vital for the resilience of both marine and freshwater environments.

Climate Regulation

Plankton are not just crucial for the survival of marine life; they also play a significant role in regulating the Earth's climate. Phytoplankton, through the process of photosynthesis, absorb vast amounts of CO_2 from the atmosphere. When phytoplankton (or organisms that have consumed phytoplankton) die, a portion of the carbon they have absorbed sinks to the ocean floor, where it can be sequestered for hundreds or even thousands of years. This process, known as the biological pump, is a critical component of the global carbon cycle, helping to keep atmospheric CO_2 levels in check, thus moderating the Earth's climate.

However, the efficiency of this carbon sequestration process is closely tied to the health of plankton populations. Changes in ocean temperatures, acidification, and nutrient availability can all impact phytoplankton growth and, by extension, the global carbon cycle. By studying plankton, scientists can better understand how these tiny organisms contribute to climate regulation and how changes in their populations might affect the Earth's climate in the future.

Water Quality Indicators

Plankton are also valuable indicators of water quality. Because they are sensitive to environmental changes, the composition of plankton communities can provide important clues about the health of aquatic ecosystems. Certain species of plankton thrive in polluted waters, while others are more common in pristine environments. By monitoring the types and abundance of plankton present in a body of water, scientists can assess the level of pollution, the presence of toxins, and the overall health of the ecosystem. For example, an increase in the population of cyanobacteria, a type of blue-green algae, often signals nutrient pollution from agricultural runoff or sewage discharge. These harmful algal blooms (Fig. 1.14) can produce toxins that are dangerous to both aquatic life and humans. On the other hand, the presence of a diverse and balanced plankton community generally indicates a healthy, well-functioning ecosystem.

In this way, plankton serve as sentinels of our waters, providing early warnings of environmental degradation. By keeping a close eye on plankton populations, we can detect and address water quality issues before they become too severe, ensuring the health of our rivers, lakes, and oceans.

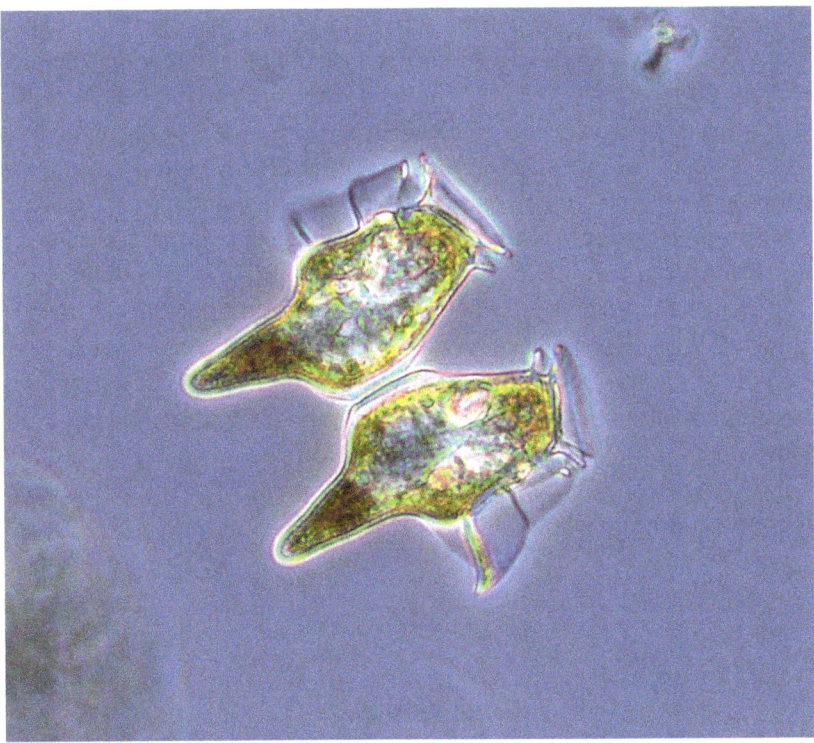

Fig. 1.14 *Dinophysis* is a toxic genus of marine dinoflagellates causing harmful blooms

Research and Education

Beyond their ecological and environmental importance, plankton are a fascinating subject of study that offers insights into some of the most fundamental questions in biology, ecology, and earth science. For researchers, plankton provide a model system for studying evolution, biogeography, and ecological interactions. Because they are found in virtually every aquatic environment on Earth, from the Poles to the tropics, plankton are key to understanding the distribution and adaptation of life across different environments.

Moreover, the study of plankton has practical applications in fields as diverse as climate science, fisheries management, alternative energies, food resources, and biotechnology. For example, research on phytoplankton has led to advancements in biofuels, as certain species are highly efficient at converting sunlight into energy-rich compounds. Similarly, understanding plankton dynamics is crucial for managing fisheries, as plankton form the base of the food chain for many commercially important fish species.

For students and hobbyists, observing plankton opens up a world of wonder. Under the microscope, plankton reveal the complexity and beauty of life at the smallest scales. This hands-on experience with plankton can inspire a lifelong interest in science and a deeper appreciation for the natural world. Through the study of plankton, we gain not only scientific knowledge but also a sense of connection to the intricate web of life that sustains our planet.

2

Getting Started

2.1 Equipment and Materials

Microscopes

A microscope is essential for observing plankton, most of which are too small to be seen with the naked eye. There are many of different kinds of microscopes:

1. **Light Microscopes:** These use visible light for illumination and are divided based on the direction of light:

   - **Compound Light Microscope:** Light is directed from *below* the specimen, passing through it. Ideal for viewing transparent or thin samples, like cells or microorganisms, mounted on glass slides.
   - **Inverted Microscope:** Light comes from above the specimen, and the objective lenses are located below the stage. These are particularly useful for observing cells or organisms in a liquid medium, like culture dishes, without the need for thin sections or slides.

© The Author(s), under exclusive license to Springer Nature Switzerland AG 2024
A. Calbet, *The Amateur Plankton Researcher's Practical Guide*,
https://doi.org/10.1007/978-3-031-80248-5_2

- **Stereomicroscope (Dissecting Microscope):** Light can be directed from above or below the specimen. It's used for observing the surface of opaque, larger, or 3D objects, providing lower magnification but good depth perception.

2. **Fluorescence Microscope:** Uses specific wavelengths of light to excite fluorescent dyes in a sample, allowing for the visualization of specific structures in color, often used for tagging proteins or DNA.
3. **Electron Microscope:**

  - **Transmission Electron Microscope (TEM):** Provides highly detailed images of internal structures at a very high resolution (up to 0.1 nm). Samples must be ultra-thin and are viewed in black and white.
  - **Scanning Electron Microscope (SEM):** Produces 3D surface images of a sample with great detail but slightly lower resolution than TEM.

4. **Confocal Microscope:** A type of fluorescence microscope that creates sharp, 3D images by focusing light on thin layers of the sample, eliminating out-of-focus light for improved resolution.

Each type serves different research needs based on the required level of detail and the nature of the sample.

Compound Light Microscopes These are ideal for viewing small, transparent plankton. They provide high magnification (typically 40x to 1000x) and are excellent for observing cellular details. We will use these microscopes to study the unicellular organisms in plankton. Under the standard magnification and resolution an amateur microscope provides, I recommend starting with a non-expensive device to get a closer look at large phytoplankton, such as diatoms and dinoflagellates, and some protozoans, such as ciliates, radiolarians, and heterotrophic dinoflagellates. If this hobby captures your interest, you might consider investing in a better microscope later on. As I said previously in the book, I started when I was about 10 years old, with a simple plastic microscope. With my first salary, I bought a better one, and so on, until I had a semi-professional model. Now, I use a professional upright microscope (the common type) to observe live samples and an inverted one (Fig. 2.1) for preserved settled samples. Both are equipped with bright field, dark field, and UV light. But I assure you, all this is not necessary to enjoy plankton samples. With a decent microscope, you can have a lot of fun. Think about Antonie van Leeuwenhoek, the discoveries he made, and the rudimentary microscope he used!

Stereomicroscopes (Dissecting Microscopes) These (Fig. 2.2) offer lower magnification (typically 10x to 40x) and are useful for larger plankton and examining samples in three dimensions. Affordable alternatives to stereomicroscopes, such as magnifying lenses for cellphones and digital enlarging microscopes (Fig. 2.3), can be found in many online and physical stores.

Features to Look For Choose a microscope with a range of magnifications to accommodate various plankton sizes. Keep in mind that no microscope will allow you to see the entire size spectrum of plankton. My advice is that if you are interested in freshwater plankton, go for a standard high-magnification microscope. However, if you are more intrigued by marine environments, start with a stereoscope to observe the larger creatures. This is because it is easier to obtain higher concentrations of protists (single-celled organisms) in freshwater than in marine environments. Go always for binocular alternatives.

Proper illumination is crucial. LED illumination is energy-efficient and provides good visibility. Some microscopes offer both transmitted (from below) and reflected (from above) light. A focus knob and adjustable stage are essential for precise viewing.

Collection Devices

Proper sampling tools and containers are necessary for collecting and transporting plankton samples.

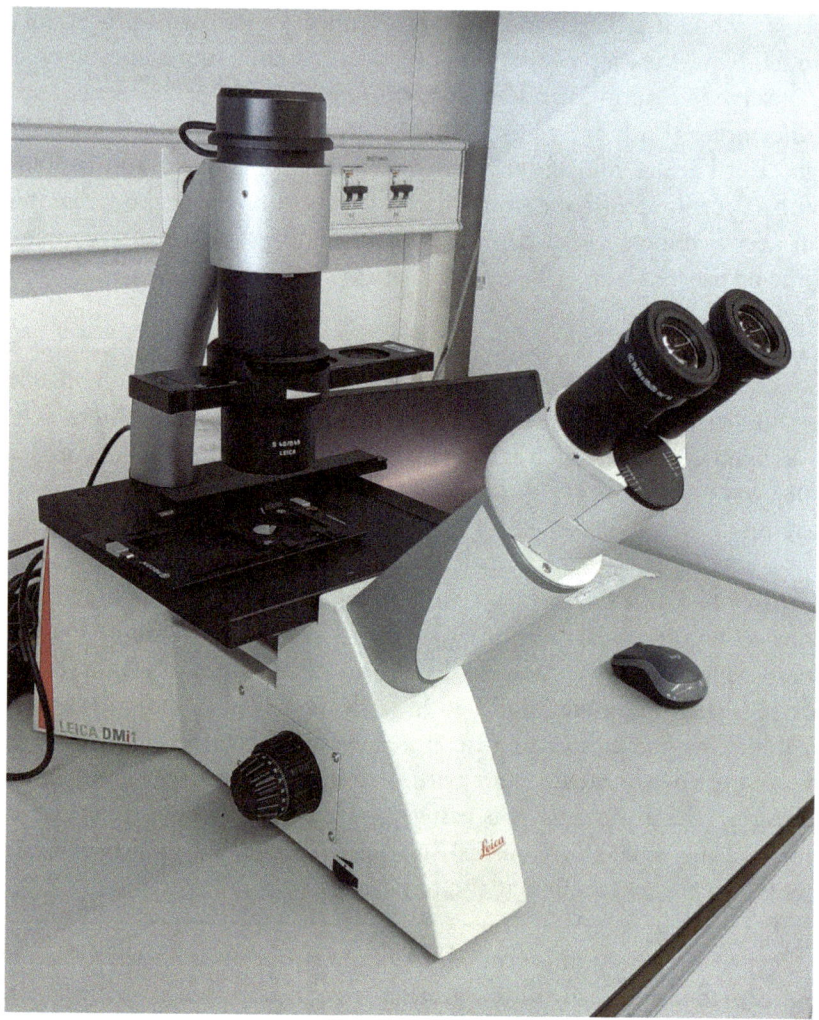

Fig. 2.1 Inverted microscope

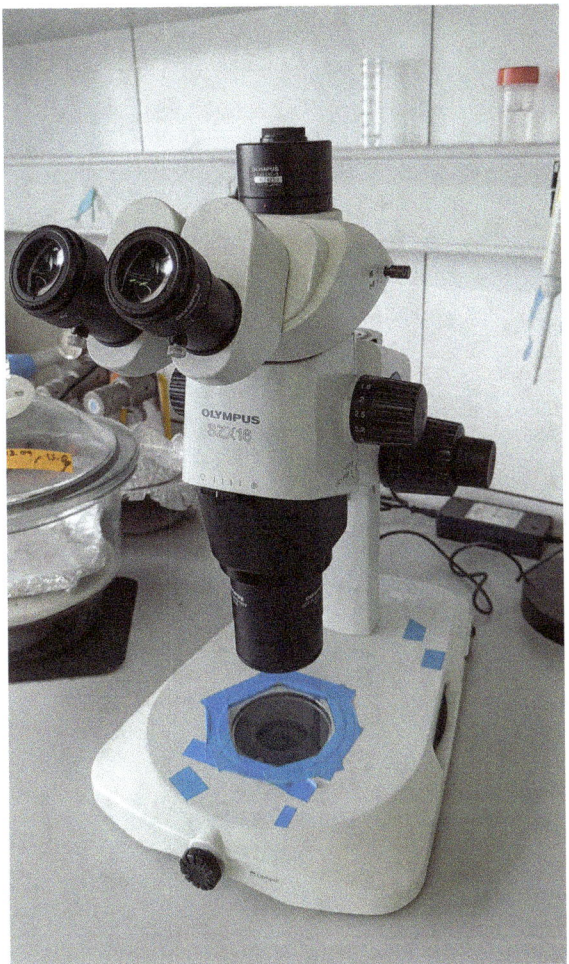

Fig. 2.2   Binocular stereomicroscope

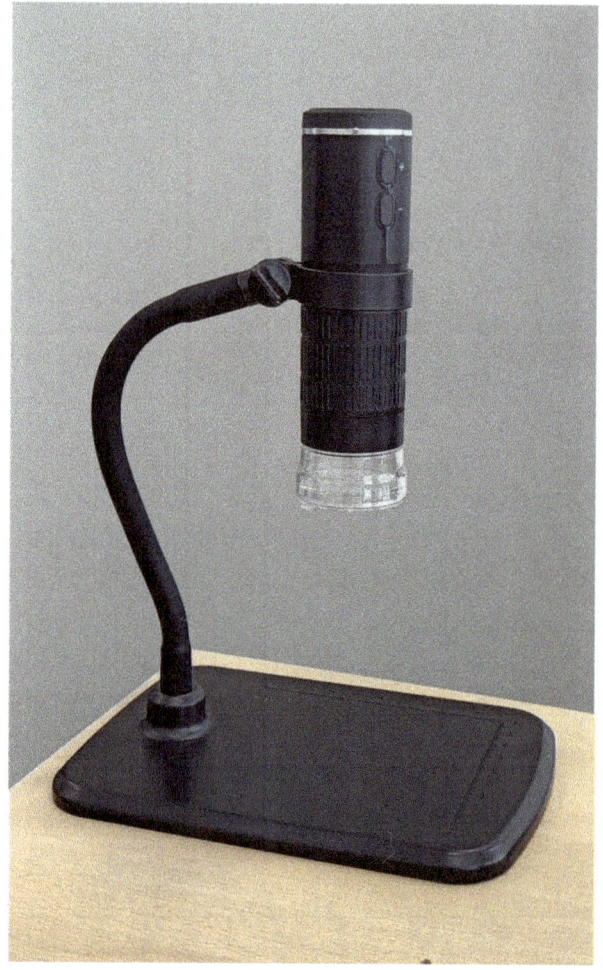

Fig. 2.3 A digital microscope that can be connected to a cellphone

- **Plankton Nets:** These are conical nets made of fine mesh, used to filter plankton from the water (Fig. 2.4). The mesh size should be appropriate for the type of plankton you are studying (e.g., 20–50 micrometers for phytoplankton and protozoans, and 100–200 micrometers for larger zooplankton). See next section, where I show you how to build your own plankton net.

Fig. 2.4 Vertical double WP2 plankton net

- **Sampling Bottles**: Use clean, wide-mouth bottles made of glass or plastic to collect and store water samples.
- **Buckets and Jars**: These are useful for larger collections or initial sample gathering before transferring to smaller containers.
- **Syringes or Pipettes**: For collecting small volumes of water and plankton from the surface or specific water layers.

How to Make Your Own Plankton Net

As you likely know, plankton is quite diluted in seawater, especially the larger kinds (zooplankton). Therefore, we use plankton nets (which can be very expensive) of varying complexity to capture these creatures. However, you can make your own plankton net (Fig. 2.5) with simple and inexpensive materials. All you need is a large plastic bottle, a nylon stocking, scissors, and some duct tape or electrical tape.

Start by cutting the bottle in half and making the largest hole possible at the base. Once you've done that, place the stocking between the two pieces of the bottle and secure it with the tape. Finally, make some small holes at the base of the bottle to tie a string so you can drag your net through the water. To help the net sink, it is useful to attach a weight to the string. The weight could be a smaller bottle filled with sand or a diving lead weight if you have one. Once you have made your catch, just open the bottle cap to collect the zooplankton in a bucket or large jar.

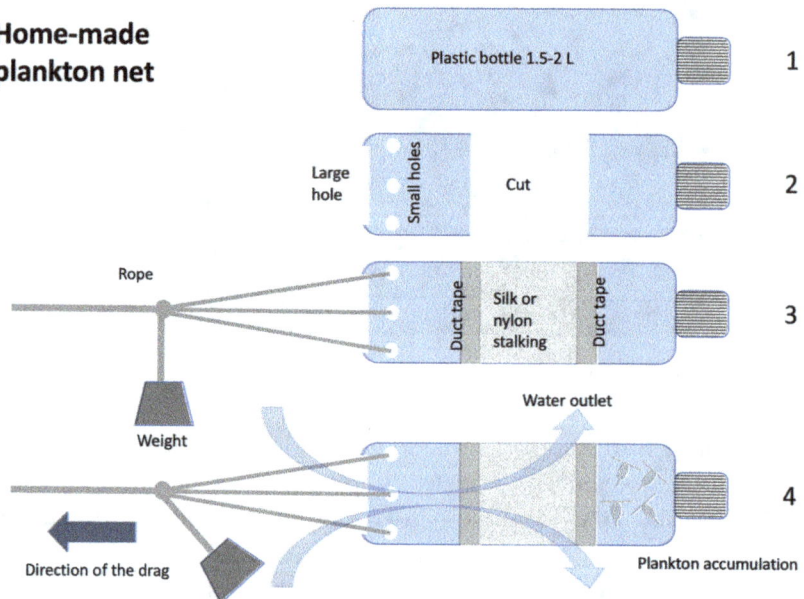

Fig. 2.5 Scheme of the construction of a homemade, inexpensive, plankton net

Preservation of the Samples

Although I recommend observing the sample fresh, preserving plankton samples is important if you cannot process them immediately (in a few hours). However, keep in mind that preserved organisms will lose their color and texture, and even can get deformed by the fixation process.

- **Preservatives**: Lugol's Solution is commonly used for preserving phytoplankton and protozooplankton; it stains cells and slows down decomposition. It can be replaced by commercial Betadine. Just add a few drops to the sample until it reaches a bourbon color. Formalin is more difficult to obtain, but it is best for metazoan zooplankton and larger organisms. The usual preservation concentration is 10% of pure formalin in the water sample. However, unless you have safety tools (gloves, a proper mask, protective glasses) and can work outside (or under an extraction hood), I would discourage the use of formalin because it is carcinogenic and highly toxic. Instead, you can use Betadine as well. Also, use it with caution due to its toxic nature, and under adult supervision if you are a child. Alcohol can be used as well, but when using denatured alcohol (the kind typically acquired at a pharmacy), the sample is likely to become cloudy and milky, which makes further observation difficult.
- **Storage Containers**: Use small vials or bottles with tight-sealing lids to store preserved samples. Label them with the date, location, depth of collection, and any other relevant information.
- **Cold Storage**: A refrigerator or cooler can be used for short-term storage of samples, especially if preservatives are not immediately available.

Identification Guides

Identifying plankton species requires reference materials. See Chaps. 7 and 9 for further details.

- **Field Guides**: Purchase or download field guides specific to your region. These guides typically include images and descriptions of common plankton species. Alternatively, I provide a short visual guide at the end of this book to help you identify the major groups.
- **Online Databases**: Websites such as AlgaeBase (www.algaebase.org), the IMOS Marine Zooplankton Guide (https://www.eoas.ubc. ca/~swaterma/473-573/Handouts/IntroductoryZooplankton FieldGuide_2014.pdf), and regional databases can provide detailed information and images. A simple search on the Internet will yield plenty of results. When starting, do not attempt to reach the species level. Give yourself some time to gain experience.
- **Mobile Apps**: Unfortunately, I am not aware of any plankton identification app. However, there are citizen science apps that can connect you with experts who will help you classify your organisms. Likewise, there are social networks of specialists and amateurs that you can join. For example, for marine water plankton, I manage a blog where you can find articles about plankton organisms and their many essential roles in Earth's ecosystems (www.planktonocean.wordpress.com).

2.2   Safety Precautions

While studying plankton at home can be safe and enjoyable, it is important to follow certain safety guidelines:

- **Handling Chemicals**: When using preservatives like formalin or iodine products, wear gloves, a mask, and protective lenses, and work in a well-ventilated area. Store chemicals out of reach of children and pets. If you are underage, always seek the assistance of an adult.
- **Microscope Safety**: It is difficult to harm yourself with a microscope (although not impossible), but it is easy to damage the microscope, particularly the delicate lenses. Handle the microscope carefully, avoiding sudden movements or impacts that could damage any of its components. Keep the microscope and lenses clean using appropriate cleaning materials, such as ethanol and specialized products.

- **Water Safety**: When collecting samples from natural water bodies, be cautious of slippery surfaces, strong currents, and potential contaminants. Wear appropriate footwear and consider bringing a friend along for safety. If sampling from a boat, do not forget your life jacket.
- **Labeling**: Always label your samples clearly to avoid confusion and ensure proper identification and handling.
- **Disposal**: Dispose of any chemical preservatives and biological waste according to local regulations. Do not pour chemicals down the drain or dispose of samples in natural water bodies. It is important not to "release" your collected plankton (if still alive) into waters different from their origin to avoid introducing invasive species.

By gathering the right equipment and following these safety precautions, you will be well-prepared to embark on your plankton study journey.

3

Collecting Plankton Samples

3.1 Choosing Collection Sites

Embarking on the adventure of collecting plankton samples is like opening a window into the hidden, bustling world of microscopic life. Selecting the right location to collect plankton samples is the first step in uncovering the rich diversity that these organisms offer. Both freshwater and marine environments present unique opportunities and challenges, and understanding these can greatly enhance the effectiveness of your sampling efforts.

Freshwater

Freshwater ecosystems, such as lakes (Fig. 3.1), rivers, and ponds (even smaller containers with accumulated rainwater), offer a variety of habitats where plankton thrive. When choosing a collection site in freshwater, consider the type of water body and its characteristics. Still, small waters, like ponds and small lakes, tend to have higher concentrations of plankton, especially in areas where sunlight penetrates the water column,

A. Calbet, *The Amateur Plankton Researcher's Practical Guide*,
https://doi.org/10.1007/978-3-031-80248-5_3

Fig. 3.1 High mountain lake. Catalonia, Spain

promoting photosynthesis. These areas are typically found near the surface and along the edges where the water is shallower.

In rivers and streams, plankton may be less concentrated due to the flow of water, but they can still be abundant in slower-moving sections, such as backwaters or near the confluence of streams. Additionally, nutrient-rich areas, such as those near inflows from agricultural land or urban runoff, may harbor higher concentrations of plankton. However, be mindful of pollution levels, as these can affect not only the quantity but also the types of plankton present, sometimes favoring harmful or less desirable species.

In freshwater environments, the best collection sites are often determined by a combination of factors, including water depth, clarity, and nutrient availability. Exploring different parts of a lake or river can reveal variations in plankton communities, offering a more comprehensive picture of the ecosystem's health and biodiversity.

For those who want a simple way to observe protozoans, you can just add some plants to a container of freshwater and watch what grows. Even a few blades of grass, leaves, or a small aquatic plant can serve as a starting point. Over time, you will notice tiny protozoans such as *Amoeba*, *Paramecium*, or *Euglena* appearing in the water, feeding on the bacteria and organic material that naturally accumulate. These microorganisms thrive in the microenvironments created by decaying plant matter. You can also try using household items like pet moss or moist soil from a potted plant as a substrate for growing protozoans. These materials often harbor dormant protozoan cysts, which will emerge when exposed to water. In just a few days, you could have a thriving mini-ecosystem full of protozoans, perfect for a low-effort exploration of the microscopic world.

Wild moss is an excellent material for observing and cultivating protozoans like amoebas. By placing moss in a container with water, you create an ideal environment for organisms such as amoebas and testate amoebas (amoebas with protective shells) to thrive, as moss retains moisture and provides a nutrient-rich habitat. Additionally, spores and other microorganisms present in the moss can serve as food for these amoebas, making it easy to observe and study them at home or in a lab setting.

Marine

Marine environments, with their vast and dynamic nature, present a distinct set of challenges and opportunities for plankton collection. The ocean is a complex system where plankton distribution is influenced by factors such as tides, currents, and temperature gradients. When selecting a marine collection site, it is essential to consider these elements to maximize your chances of capturing a diverse plankton sample.

Coastal areas, where nutrient-rich waters from rivers and upwellings mix with the ocean, are often prime locations for plankton sampling. These zones, known as estuaries (Fig. 3.2) or coastal upwellings, are nutrient-rich and teeming with life, making them ideal for collecting a wide range of plankton species. Areas near coral reefs, seagrass beds, and kelp forests can be particularly fruitful, as these ecosystems support high plankton productivity. Do not disregard sportive and commercial harbors; even if polluted, they often hold a large variety of organisms.

When venturing further offshore, plankton can be found at various depths, depending on factors such as light availability and water temperature. Surface waters, especially in regions with strong sunlight, are often rich in phytoplankton, the primary producers that form the base of the

Fig. 3.2 Estuaries are often good places for collecting plankton. Delta Ebro Estuary, Catalonia, Spain

marine food web. However, sampling at different depths can reveal different layers of the plankton community, including zooplankton that feed on the phytoplankton below.

Marine plankton collection often requires careful planning, considering tides, currents, and weather conditions, all of which can influence the success of your sampling efforts. By selecting sites with known productivity, such as upwelling zones or nutrient-rich coastal waters, you increase your chances of collecting a representative sample of the marine plankton community.

If marine waters are unavailable for any reason, you can still observe some planktonic protozoans by examining the water found inside mussels and other bivalves under a microscope. Bivalves, such as mussels, clams, and oysters, filter large amounts of water through their bodies, trapping plankton, protozoans, and other microorganisms in the process. By carefully extracting the water from their internal cavities or siphon areas, you can find a rich variety of microscopic life, including ciliates and flagellates. These organisms thrive in the nutrient-rich environment provided by the bivalve's filtering activity, and studying them can give you a fascinating glimpse into the microscopic world of marine protozoans without needing direct access to seawater.

3.2   Time of the Day

The time of day can significantly impact plankton sampling, as many plankton species exhibit diel vertical migration, moving up and down the water column in response to changing light conditions. Understanding this behavior is crucial to timing your sampling efforts effectively.

During the day, many zooplankton species descend to deeper, darker waters to avoid predators and return to the surface at night to feed on phytoplankton. This nocturnal migration means that nighttime sampling can yield different results from daytime sampling, often with higher concentrations of zooplankton near the surface. Phytoplankton and proto-zooplankton, on the other hand, mostly remain in the upper layers of the water column where sunlight is available for photosynthesis. Therefore, if your goal is to collect phytoplankton, daytime sampling is generally effective.

In freshwater environments, the time of day can also influence plankton distribution. Morning sampling for phytoplankton might reveal higher concentrations of plankton near the surface as they rise to take advantage of the early sunlight. In contrast, zooplankton, similar to those in marine systems, should be sampled at night or during the day closer to the bottom for better results.

In marine environments, tides and currents can further complicate the timing of your sampling efforts. Sampling during a rising tide might bring in plankton from offshore, while a falling tide could draw plankton from coastal areas. Similarly, sampling during different phases of the moon can affect plankton behavior, particularly in species that are sensitive to light.

In summary, the art of collecting plankton samples lies in understanding the environment you are working in and making informed decisions about where and when to collect. Whether you are exploring the still waters of a freshwater lake or the dynamic coastal zones of the ocean, the choice of site and timing can significantly impact the diversity and abundance of plankton in your samples. By carefully selecting your collection sites and considering the time of day, you can unlock the secrets of the microscopic world and gain a deeper appreciation for the complex and vital role plankton play in our planet's ecosystems.

4

Observing Plankton

4.1 Preparing Samples for Observation

Proper preparation of your plankton samples is crucial for successful observation under the microscope. Follow these steps:

Collecting Fresh Samples

Whenever possible, use freshly collected samples for observation. Fresh samples provide the best representation of living plankton. Observing a preserved sample can be disappointing in comparison with a fresh one (Fig. 4.1). The organisms not only lose their motility, but their colors also fade, making them translucent instead of transparent. This effect is particularly noticeable when observing protists; apart from diatoms, most soft-bodied protists will simply burst upon preservation, and those that maintain their integrity will appear with unnatural shapes and colors.

© The Author(s), under exclusive license to Springer Nature Switzerland AG 2024
A. Calbet, *The Amateur Plankton Researcher's Practical Guide*,
https://doi.org/10.1007/978-3-031-80248-5_4

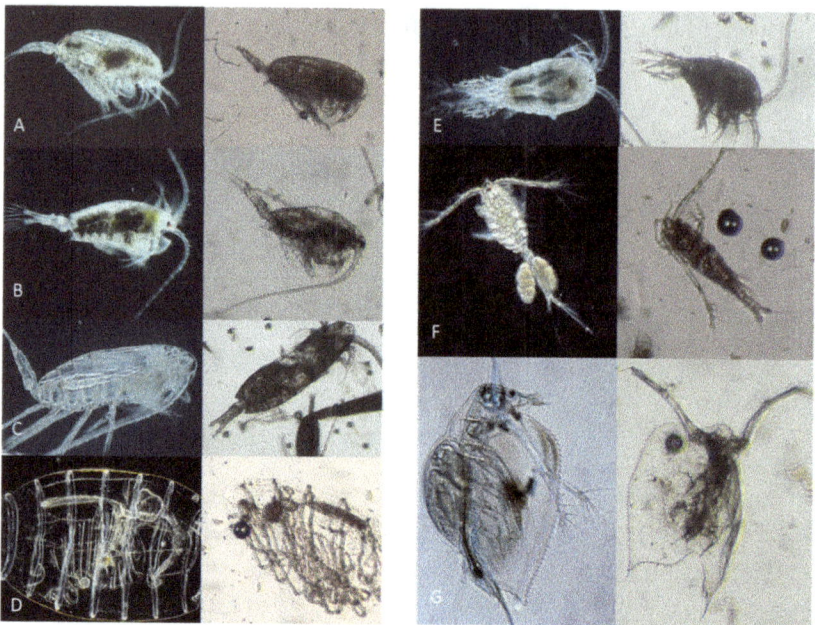

Fig. 4.1 Live (left) *vs* formalin-preserved (right) zooplankton. (**a**) *Clausocalanus* sp. (**b**) *Centropages typicus,* (**c**) *Calanus* sp., (**d**) *Doliolum nationalis,* (**e**) *Temora stylifera,* (**f**) *Oithona* sp., (**g**) *Penilia avirostris*

Concentrating Plankton

If your sample is too dilute, you can concentrate it by allowing it to settle for several hours or even overnight (though this method will not work for actively swimming organisms). The heavier organisms will naturally sink to the bottom, and you can then carefully siphon off the top water and examine the concentrated layer. Alternatively, you can use a plankton net or even a small sieve to filter out larger organisms, obtaining a more concentrated sample for observation. For protozoans that seem too diluted to observe well, I recommend splitting your sample into different containers. This is particularly helpful if you are more interested in observing a range of organisms rather than focusing on exact abundance measurements.

Once divided, you can experiment by adding varying concentrations of nutrients to each container to stimulate the growth of microorganisms. A few drops of plant fertilizer—which contains key nutrients like nitrogen and phosphorus—work well for this purpose. Avoid adding too much, as high nutrient concentrations can cause the water to become eutrophic and kill sensitive organisms. Let the samples rest for a couple of days near a well-lit window, but be sure to avoid direct sunlight, which can cause overheating. After a few days, a new plankton community should emerge, giving you a richer sample to observe under the microscope. This method provides an easy way to cultivate protists at home or in a classroom setting and allows you to explore their dynamic growth and interactions.

Preparing Slides

For macroscopic organisms visible with a stereomicroscope, a petri dish or a similar container will suffice. However, for microscopic observation, you will need to prepare slides. Place a drop of your sample on a clean glass slide and cover it with a cover slip, taking care to avoid air bubbles. If you wish to observe a bit more of the sample, you can use excavated slides (Fig. 4.2). If these are not available, you can slightly melt the edges of the coverslip using a flame (Fig. 4.3). Handle the coverslip with tweezers and briefly expose the edges to the flame. This will create a small bending of the edge, allowing you to add an additional volume of water between the cover slip and the slide. When observing live organisms, particularly protists, their rapid movement can be frustrating. In these cases, consider using methods to slow down the plankton. An old technique involves adding agar to the sample. Another method uses tobacco smoke or ether vapors. You may also discover a new technique that works best for you. Good luck!

Fig. 4.2 Normal (top) and excavated (bottom) slide

Fig. 4.3 Process of modifying a cover slip to allow more volume inside

4.2 Using a Microscope

Setting Up the Microscope

- Placement: Set up your microscope on a stable, flat surface in a well-lit area.

- Lighting: Ensure the microscope's light source is functioning correctly. Adjust the brightness to a comfortable level. For those microscopes with diaphragm in the light, close it as much as possible according to your magnification; you should not see dark edges in your field of view, but it should be close to it.
- Condenser: For microscopes with an adjustable condenser, position the condenser so that it is centered under the objective lens. Use the condenser focus knob to raise or lower the condenser to its optimal position for your objective lens. Adjust the condenser diaphragm to control the amount of light and the contrast of the image. Finally, ensure the light is evenly distributed across the field of view for the clearest image.
- Cleaning: Clean the lenses using lens paper or a soft cloth. Avoid touching the lenses with your fingers.

Adjusting Magnification

- Starting Low: Begin with the lowest magnification (e.g., 4x or 10x) to locate your sample. This provides a broad field of view.
- Increasing Magnification: Once you have located your sample, gradually increase the magnification by switching to higher power objectives (e.g., 40x, 100x). Because most amateur microscopes are not perfectly parfocal, you should refocus the image after changing magnification.
- Oil Immersion (if available): For very high magnification (e.g., 1000x), you may need to use an oil immersion lens. **Not all objectives can be immersed in oil**, so please check the details on your objective; it should be labeled "oil." Place a drop of microscopy immersion oil on the cover slip and carefully lower the objective into the oil.

Focusing Techniques

- Coarse Focus: Use the coarse focus knob to bring the sample into general focus when using low magnification.
- Fine Focus: Switch to the fine focus knob for precise focusing, especially at higher magnifications.

- Adjusting the Stage: Move the stage gently to center different parts of the sample in the field of view. Most microscopes have stage controls to move the slide horizontally and vertically.
- Depth of Field: Adjust the focus to view different layers of the sample. Plankton can be three-dimensional, and different structures may come into view as you adjust the focus. By closing the condenser diaphragm, you can achieve a greater depth of field, but this may reduce clarity.

Taking Photographs

Photographs are a powerful tool for documenting and sharing your findings.

- Microscope Cameras: If your microscope has a built-in camera or can accommodate one, use it to capture detailed images of your samples. Ensure the camera is properly aligned with the eyepiece.
- Smartphone Photography: You can use your smartphone camera through the microscope's eyepiece. Use an adapter designed for this purpose to stabilize the phone and improve image quality.
- Lighting and Focus: Adjust the lighting and focus to obtain the clearest image possible. Use the microscope's fine focus knob to ensure the subject is sharp.
- Image Editing: You may use basic photo editing software to enhance the clarity, brightness, and contrast of your images if necessary. Avoid over-editing, as this can distort the actual appearance of the plankton.
- Labeling and Storing: Organize your photos in folders named by date and sample location. Label each photo with relevant details for easy reference.

4.3   Identifying Common Plankton Species

At the end of this book, you will find an image library of the most common groups of plankton to aid in identification.

Phytoplankton are plant-like plankton that perform photosynthesis. They can be identified by their unique shapes and structures.

- Diatoms: Characterized by their silica cell walls that form intricate patterns. They come in various shapes, including pennate (elongated) and centric (circular). Many times they form long chains.
- Dinoflagellates: Identified by their two flagella, which aid in movement. They often have a cellulose armor called theca, with distinctive shapes and grooves.
- Cyanobacteria: Also known as blue-green algae, their filamentous or colonial forms can serve to identify these. They are often blue-green.
- Green Algae: These come in various forms, from single cells to colonies. They contain chlorophyll, giving them a green color.

Zooplankton are animal-like plankton that feed on other plankton or organic material. They can be identified by their movement and body structures.

- Protozoans: Single-celled organisms with various shapes. Examples include amoebas (with pseudopodia), heterotrophic dinoflagellates, and ciliates (covered with cilia for movement). A particular group of ciliates, tintinnids, possess a rigid cup-shaped structure called lorica.
- Copepods: Small crustaceans with segmented bodies and long antennae. They move with a jerky motion. The young larval stages of copepods (nauplii) look very different from copepodites (older larvae) and adults.
- Cladocerans: Also known as water fleas, they have a distinct carapace and move by hopping. They are abundant in freshwater and coastal marine systems during the summer.
- Ostracods: Small crustaceans, particularly abundant in freshwater. They possess a carapace with two shells resembling a bean.
- Rotifers: Identified by their wheel-like cilia at the front, which they use for feeding and locomotion. Do not mistake them for unicellular ciliates.

By carefully preparing your samples, correctly setting up and using your microscope, and familiarizing yourself with common plankton species, you will be well-equipped to observe and identify these fascinating organisms.

5

Conducting Experiments

5.1 Designing Simple Experiments at Home

Conducting experiments is an exciting way to explore the behaviors and ecological roles of plankton. This chapter will guide you through designing simple experiments, focusing on the effects of light and nutrients on plankton growth, and the methods for recording and analyzing data. To design effective experiments (Fig. 5.1), it is important to formulate a clear hypothesis, control the variables, and ensure repeatability. Here, we outline a few straightforward experiments you can perform at home.

Experiment 1: Observing the Effect of Light on Plankton Movement

Aim

To observe how plankton responds to different light conditions and understand their phototactic behavior (movement in response to light).

A. Calbet, *The Amateur Plankton Researcher's Practical Guide*,
https://doi.org/10.1007/978-3-031-80248-5_5

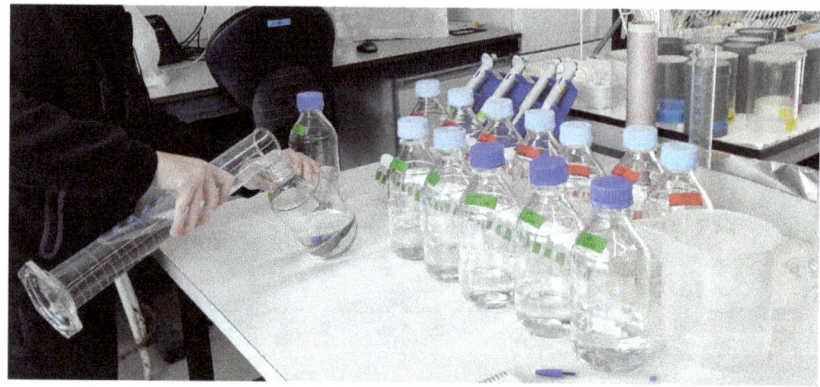

Fig. 5.1 Experimental setup

Materials Needed

- A jar or a clear container
- Pond water or seawater (make sure it contains plankton; you can collect it from a local pond or beach)
- A flashlight or a desk lamp
- Aluminum foil or black paper
- A timer or watch
- Notebook and pen for observations
- A magnifying glass or smartphone with a magnifying app (optional, but helpful)

Procedure

1. Collect Plankton: Fill a clear jar or container with pond water or seawater. Collect it from a location with visible plankton. The water should be relatively clear with some small particles visible.
2. Prepare the Experiment Setup: Wrap one side of the jar with aluminum foil or black construction paper. This will create a shaded area within the jar, leaving the other side exposed to light. Alternatively, you can create a gradient environment by covering the bottom portion of the jar, leaving the upper area exposed to light. This mimics

natural conditions where light penetrates the surface while deeper areas remain shaded, promoting different types of microorganisms based on their light preferences.

3. Introduce the Light Source: Place the jar on a flat surface. Use the flashlight or desk lamp to shine light on the exposed side of the jar. Make sure the light source is steady and consistently shines on one side only.

4. Observation: Start the timer and observe the jar. Look closely at the water and see if the plankton start moving toward or away from the light. Use the magnifying glass or smartphone to help you see the plankton better, if necessary. Record your observations every 5 minutes for a total of 20–30 minutes.

5. Switch Light Conditions (Optional): After 20–30 minutes, you can switch off the light or move it to the other side of the jar to see if the plankton change direction.

6. Conclude the Experiment: After completing your observations, compare the behavior of the plankton under different light conditions. Did they move toward the light (positive phototaxis), or away from it (negative phototaxis)? With a Pasteur pipette you can even sample the concentrated patches of plankton to observe what species are dominant.

7. Record and Discuss: Write down your observations, including the time it took for the plankton to respond to the light, the direction they moved, and any other interesting behavior you noticed. Discuss why you think the plankton behaved the way they did. Consider how light might affect plankton in their natural environment.

Expected Results

Plankton are often phototactic, meaning they move in response to light. You may observe that most plankton move toward the light (positive phototaxis) because they rely on light for photosynthesis (if they are phytoplankton) or to find food. Some plankton may move away from the light (negative phototaxis) as a way to avoid predators or harmful UV rays.

Conclusion

This experiment helps you understand how light influences plankton behavior, a key aspect of their survival in aquatic ecosystems. It also demonstrates the importance of light in the distribution and movement of plankton in their natural habitats.

Experiment 2: Studying Plankton's Response to Water Acidity (pH)

Aim

To observe how changes in water acidity (pH) affect plankton behavior and survival.

Materials Needed

- Three clear jars or containers
- Pond water or seawater with plankton
- Vinegar (to lower pH)
- Baking soda (to raise pH)
- pH test strips or a digital pH meter (optional but recommended)
- Notebook and pen for observations
- Optional: A magnifying glass or smartphone with a magnifying app

Procedure

1. Collect Plankton: Fill all three jars with the collected pond water or seawater, ensuring there is plankton in each jar. I would advise filling only one-third of each jar at a time to homogenize the plankton in all of them. If you want to conduct a more sound and realistic experiment, you should do replicates (at least three) for each treatment.
2. Label the Jars: Label the jars as "Acidic," "Control," and "Basic."

3. Adjust the pH Levels: The "Control" jar should remain untreated, maintaining the natural pH of the pond or seawater. Add a small amount of vinegar to the "Acidic" jar, stirring gently. Use pH strips or a digital pH meter to lower the pH to around 4–5. Dissolve a small amount of baking soda in the "Basic" jar, raising the pH to around 9–10. Note that for seawater you should add more vinegar or baking soda than in freshwater. This is because of the pH-dampening effects of sea salts.

4. Observation: Observe the plankton in each jar over several hours or days. Note their movement, survival, and any changes in water clarity or color. Record your observations regularly, paying attention to any differences between the jars.

5. Analyze the Results: Compare the plankton in the three different pH conditions. Which condition seems to support plankton survival best? Are there noticeable differences in their activity?

Expected Results

Plankton may struggle in the "Acidic" jar as many aquatic organisms are sensitive to low pH levels, which can disrupt their cellular processes. The "Basic" jar may also show reduced plankton activity compared to the "Neutral" jar.

Conclusion

This experiment highlights how water acidity affects plankton, demonstrating the potential impact of environmental changes, such as acid rain or ocean acidification, on aquatic life. These two experiments are very simple, engaging, and provide valuable insights into the behavior of microscopic organisms that play a crucial role in aquatic ecosystems. However, for a more professional experiment see the next experiments.

Experiment 3: Light Effects on Plankton Growth

Aim

To investigate how different light intensities affect phytoplankton growth. We hypothesize that the more light provided the faster the growth of phytoplankton, up to a level in which phytoplankton (Fig. 5.2) would not grow more and even will start dying.

Materials

- Clear containers (e.g., glass jars or transparent plastic bottles)
- Phytoplankton culture (can be obtained from a local pond or a biological supply company. See the cultivating plankton section at the end of the chapter). Alternatively, you may use the entire plankton community using a fresh sample. In this way, you will see the global effects of light on the community.
- Light sources (e.g., sunlight, LED lamps, etc.)

Fig. 5.2 Phytoplankton culture of *Rhodomonas salina*. Note that this particular microalgae is red instead of green

- Light intensity meter (optional, for precise measurement)
- Algae growth medium (commercially available or homemade)
- Measuring spoons and cups
- Labels and markers

Procedure

1. Prepare the algae growth medium according to instructions and distribute it equally among the clear containers. Use, at least, triplicated containers per treatment. The containers should be clean (it is better if you boil them to obtain sterile conditions). It is important for this experiment that the growth medium is the same in all bottles to avoid adding a confounding factor.
2. Inoculate each container with a small amount of the phytoplankton culture.
3. Label the containers with different light intensity levels (e.g., low, medium, high).
4. Place the containers in different locations with varying light intensities or use different light sources positioned at varying distances[1]. A way of getting a gradient of light is placing all containers under the same light intensity and covering the different treatments with plastic bags that allow different light penetrations. The temperature of all treatments must be the same.
5. If possible, measure the light intensity with a light meter and note the values. A simple photometer from a camera or cell phone will do it.
6. Allow the experiment to run for a set period (e.g., one week), maintaining consistent light exposure for each container.
7. Observe and record changes in the color and density of the phytoplankton daily. You can take pictures of the progression.

---

[1] An alternative to this experiment could be to cover the flask with different color semi-transparent layers. In this way, you will test how the light spectrum (color) would affect phytoplankton growth.

Data Collection

8. Ideally, measure the growth of phytoplankton using a spectropho-tometer if available (I guess not), or visually estimate the density by comparing the color intensity of the cultures.

9. A more precise, but laborious way, would be taking samples of each container, after gently mixing, and preserving them with iodine solu-tion. Afterward, under the microscope, you can count the number of cells present in each treatment. To do so, you will need to homoge-neously distribute a drop of water sample under the cover slide and use the most adequate magnification to count the cells in each view field. For consistent results, you will need to count several fields (20-30 fields) of each replicated sample. To translate your cell counts into real abundance you would need to determine the area of each of your vision fields and the area and volume under the slide. The easi-est way would be always dropping a known volume of water (e.g., 0.5 milliliters) into the slide (this can be done with a graduated plas-tic Pasteur pipette) and measuring the area of your cover slide. Now, using a ruler under the microscope to know the diameter of your field of view and with the help of some simple mathematics, you can approximate the area of your field of vision. Knowing all these vari-ables you can determine the actual concentration of prey with a sim-ple rule of three. A more precise (although expensive) alternative would be acquiring a 1 milliliter Sedgewick Rafter counting cham-ber (Fig. 5.3), which has squares of known volume. By counting the cells in each little square under the microscope you can make a con-version to the actual concentration in cells/ml.

10. Record the observations systematically.

Analysis

11. Compare the growth of phytoplankton under different light intensi-ties. To obtain phytoplankton growth you need to apply the follow-ing equation:

Fig. 5.3   Sedgewick Rafter counting chamber

12. $\mu = \ln(N2/N1)/(t2 - t1)$, where μ is the specific growth rate of the phytoplankton algae, and $N1$ and $N2$ are the abundance at time 1 ($t1$) and time 2 ($t2$), respectively. The time should be in hours or days. Graph the growth trends to identify the light intensity that promotes the highest growth rate.

13. You can perform a simple statistical test (such as the Student's t-test) when there are only two groups of samples to compare and the distribution of the data follows a Normal curve (bell-shaped). When there are more than two groups or factors involved, you should conduct ANOVA or other statistical tests. Given the nature of this particular experiment, you may want to see if the data adjust to a regression line (i.e., when plotting the growth of the algae as a function of the light intensity). Keep in mind that without replication of the treatments, no statistical test is possible.

Experiment 4: Nutrient Effects on Plankton

Aim

To explore how varying nutrient concentrations affect plankton growth. In this case, we want to keep the light intensity and temperature constant.

Materials

- Clear containers (e.g., glass jars or plastic bottles)
- Phytoplankton culture or natural community of plankton.
- Different nutrient solutions (e.g., varying concentrations of nitrogen or phosphorus) or a gradient of a plant fertilizer including all nutrients.
- Measuring spoons and cups
- Labels and markers

Procedure

1. Prepare different nutrient solutions with varying concentrations (e.g., low, medium, high) using commercially available fertilizers or home-made mixtures using plant fertilizers.
2. Distribute the solutions among the clear containers.
3. Inoculate each container with a small amount of the plankton culture (always the same and remember to mix well, but gently, the culture before taking your aliquot).
4. Label the containers with the nutrient concentrations.
5. Place the containers in a location with consistent light and temperature conditions.
6. Allow the experiment to run for a set period (e.g., one week).
7. Observe and record changes in the color and density of the plankton daily.

Data Collection

8. Measure plankton growth as before.
9. Record the observations systematically.
10. Analysis:
11. Compare the growth of plankton under different nutrient concentrations.
12. Graph the growth trends to identify the nutrient concentration that promotes the highest growth rate.

Experiment 5: Bringing Resting Stages to Life

For those who are curious about exploring the hidden world of plankton resting stages, a simple experiment can offer fascinating results. All that is required is some sediment, a container, and a bit of patience. By collecting mud from the bottom of a pond, lake, or coastal area, you can create a small-scale environment in which resting stages can hatch and come to life.

1. **Collect Sediment:** To begin, carefully gather a sample of mud from the sediment layer of a body of water. Ideally, this sample should come from an area that has not been disturbed recently, as this will increase the likelihood of containing viable resting stages.
2. **Prepare the Environment:** Place the mud in a clear container, such as a glass jar, and fill it with filtered water from the same location where the mud was collected (to avoid contamination I would recommend making it reach 100 °C and cooling it down again). Ensure that the container has access to light, as many plankton species require light to emerge from dormancy.
3. **Wait for Emergence:** Place the jar in a location where it will receive moderate warmth and light, and then wait. Over time, if resting stages are present, you may begin to see small planktonic organisms hatching and swimming in the water. This process can take several days to weeks, depending on the species and the conditions. This is in essence the method used to hatch *Artemia salina* eggs to feed fish.

4. **Observe and Record**: As the resting stages hatch, observe the changes in the water. Use a magnifying glass or a microscope to examine the tiny organisms that emerge. By documenting your observations, you can begin to understand the diversity and life cycles of the plankton that were previously hidden in the sediments.

5.2   Recording and Analyzing Data

Recording and analyzing data systematically is crucial for drawing valid conclusions from your experiments.

Recording Data

- Data Sheets: Use standardized data sheets, such as Excel files, to record daily observations. Include columns for date, light intensity (or nutrient concentration), plankton density, and any additional notes.
- Photographs: Take photographs of the cultures at regular intervals to visually document changes over time.
- Consistent Timing: Make observations at the same time each day to minimize variations due to daily light cycles or temperature fluctuations.

Analyzing Data

- Graphing Results: Use graphing software or draw graphs by hand to visualize the data (Fig. 5.4). Plot light intensity (or nutrient concentration) on the x-axis and plankton growth on the y-axis.
- Statistical Analysis: If you have multiple replicates, calculate averages and standard deviations to assess the reliability of your data.
- Trend Analysis: Look for trends or patterns in the data. Determine which light intensity or nutrient concentration had the most significant impact on plankton growth.
- Conclusion: Based on your analysis, draw conclusions about the effects of light and nutrients on plankton growth. Discuss any potential experimental errors and suggest improvements for future studies.

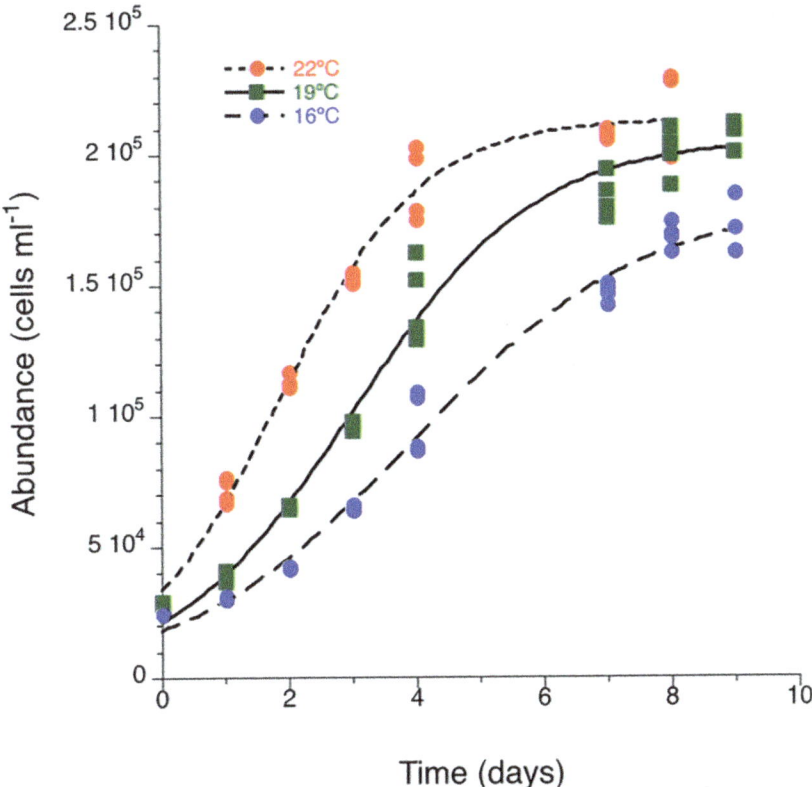

Fig. 5.4 Abundance over time of three cultures of the diatom *Thalassiosira weiss-flogii* grown at three different temperatures

By designing thoughtful experiments, carefully recording observations, and analyzing data systematically, you can gain valuable insights into the factors that influence plankton growth and contribute to our understanding of plankton ecology.

From here, you may complicate your experiments by involving, for instance, zooplankton as well. You can capture the zooplankton with a wide-mouth Pasteur pipette, one by one, and place it into the experimental bottles. Given zooplankton prey on phytoplankton, you should set up controls with only microalgae to account for the growth of the prey. These controls should have the exact conditions of the experimental bottles but obviously should not contain zooplankton. The calculations to obtain grazing rates are quite complex and require a bit more effort and mathematical skills. Therefore, I would not include them here; you can find them with a simple search on the internet. Yet, you can compare the color of experimental and control bottles to observe differences. Ideally, where you added zooplankton, the color should be paler compared to the controls, where phytoplankton could grow without grazing pressure.

5.3   Cultivating Plankton at Home

Cultivating single species of phytoplankton and zooplankton at home is a straightforward process for some species, although very complicated and frustrating for others. The key to success lies in creating and maintaining the right conditions for growth, which include providing the correct food, light, and oxygen.

Keep in mind that the most axenic conditions you may provide the best. Meaning that you will need to either boil or microwave all the material. To cultivate phytoplankton, you will need a clear container, such as a glass jar or plastic bottle, ranging from 10 to 100 milliliters in size (let's start small). The first step is to prepare the water. If you are cultivating marine phytoplankton, it is best that you get local marine water and filter it through a coffee filter. Then you may try to kill all the remaining bacteria by microwaving it up to start boiling (do not let it boil), then stop, and do it again, several times for a couple of minutes (switching on and off the microwave), and then cool it down fast by introducing the bottle into icy water. Use a bottle that can resist these abrupt temperature changes (e.g., Pyrex glass). Boiling marine water for some time usually produces the precipitation of carbonates and other essential salts, which is why I recommend not reaching 100 °C for long. An alternative would be you add marine salt (from an aquarium store, not table salt) to sterilized distilled water until the salinity reaches between 30 and 35 (i.e., 35 grams of salt per liter of water), although this procedure is not suitable for all species. For freshwater phytoplankton, use bottled water (not from the tap) previously sterilized. Once the water has cooled down, add a nutrient medium designed for phytoplankton growth, such as Guillards F/2, or a commercially available fertilizer diluted accordingly. This medium should provide essential nutrients like nitrogen, phosphorus, and trace minerals that phytoplankton need to thrive. A rough approximation would be to add about 1 milliliter of this nutrient solution per liter of water. But this will largely depend on the fertilizer you use. For diatoms, you will need to add silicate as well, which is difficult to get, sorry.

Next, introduce the phytoplankton starter culture to the prepared water. A good rule of thumb is to use a 10% inoculum, meaning you add 10 milliliters of the starter culture to 90 milliliters of water. The tricky question is how to obtain this original culture. There are several options. You may buy it through the Internet, you may ask for it from a known scientist who is willing to share the cultures with you, or you may try to isolate it. Isolating your own culture is quite fun, although it is not easy. To do so, you will need to isolate one by one single cells of the selected species and place each of them on an independent little vial (approx. 5 mL) filled with medium suspension. I use a multiwell plate (Fig. 5.5), but these are not easy to get outside the laboratory. To individually collect the algae, you will need to fabricate a very thin glass Pasteur pipette. This can be easily done by holding the tip of the pipette with metallic tweezers over a flame and pulling it when starts melting (Fig. 5.6). In this way, the tip will stretch and become thinner. You may need to repeat the process several times and waste a bunch of pipettes. But, for sure, in the end, you will be satisfied with one of them.

Whenever you are sure one of the desired cells is inside each vial, close them and let them rest in an illuminated area (avoid direct sunlight) at room temperature. After a few days, or even weeks, in some of the containers you will see that algae are growing. Check the species present under the microscope and use the ones with your selected species only.

Once the culture is added, place the container under a full-spectrum light source, such as fluorescent or LED lights, that mimics natural sunlight. You may use sunlight, but avoid overheating of the culture or too much intensity. Phytoplankton require 12 to 16 hours of light daily to photosynthesize and grow.

Over the next few days, you should see the water gradually turn greener (if the microalgae are green), indicating that the phytoplankton is

Fig. 5.5 Multiwell plate

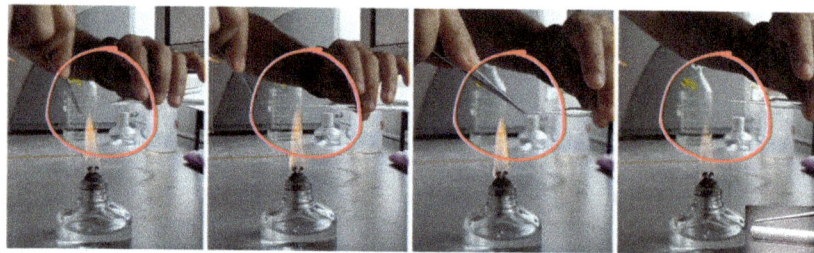

Fig. 5.6 Details on the process of preparing a thin tip Pasteur pipette

multiplying. The culture will typically reach its peak density in about one to two weeks. At this point, you can harvest the phytoplankton by removing a portion of the culture and replacing it with fresh, nutrient-enriched water. The harvested phytoplankton can be fed to zooplankton or used for experiments.

For protozoans, particularly the freshwater ones, I recommend adding a couple of grains of rice or wheat to the water to obtain dense cultures of bacterivorous ciliates or similar. Once you have enough organisms you may decide to isolate them one by one, like for microalgae, and feed them with previously boiled grains of rice and wheat.

For larger zooplankton cultivation, start with a similar container, but bigger, filled with either marine or freshwater, depending on the species of zooplankton you are growing. Although the water should be previously filtered, the sterility conditions are not so important here. Common zooplankton species include rotifers, copepods, and cladocerans (freshwater ones), all of which are relatively easy to culture at home. They can be captured with a wide-mouth Pasteur pipette. Begin by introducing a small starter culture of zooplankton into the prepared water. Zooplankton feed on phytoplankton, so you will need to add a portion of your phytoplankton culture to the zooplankton container every day. The quantity of phytoplankton added should be enough to keep the water slightly green, indicating that there is enough food for the zooplankton but not so much that it clouds the water and decreases oxygen levels. In the case of freshwater cladocerans or rotifers, there are alternative dried yeast, chlorella, or spirulina that may work as well. At least for some time. The problem is that these dried yeast and algae usually make the water dirty very fast, and anoxia events can readily occur.

Zooplankton cultures require gentle aeration to keep the water well-oxygenated. It is also important to maintain a stable temperature, usually between 16 °C and 25 °C (depending on the temperature at the collection time), and avoid sudden changes that could stress the zooplankton. Over time, the zooplankton population will grow, and you can begin harvesting them as needed. Simply siphon out a portion of the zooplankton-rich water and replace it with fresh water. The harvested zooplankton can be used as a nutritious food source for aquarium fish or other marine and freshwater organisms.

6

The Role of Molecular Techniques in Plankton Identification

About This Chapter

I added this chapter, not for you amateur planktologist to make use of it, but to keep you informed about the current methods to identify plankton, their advantages, and caveats. Traditional methods of identifying plankton, which rely on visual observations under a microscope, are time-consuming and, more importantly, require expert knowledge. However, advances in molecular techniques have revolutionized our ability to identify and study these organisms with unprecedented efficiency. These techniques, which include DNA barcoding, metagenomics, and environmental DNA (eDNA) analysis, are unlocking new insights into the hidden world of plankton and transforming the field of marine biology.

6.1 DNA Barcoding: The Genetic Fingerprint of Plankton

One of the most powerful tools in molecular plankton identification is DNA barcoding. This technique involves sequencing a short, standardized region of DNA from an organism and comparing it to a reference database to identify the species. In the case of plankton, DNA barcoding

© The Author(s), under exclusive license to Springer Nature Switzerland AG 2024
A. Calbet, *The Amateur Plankton Researcher's Practical Guide*,
https://doi.org/10.1007/978-3-031-80248-5_6

can be used to identify even the tiniest organisms that are otherwise indistinguishable under a microscope.

The process begins by extracting DNA from a plankton sample, which may contain hundreds or thousands of different species. A specific gene, often the mitochondrial cytochrome c oxidase I (COI) gene, is then amplified using polymerase chain reaction (PCR) and sequenced. The resulting DNA sequence is compared to a database of known sequences to determine the species present in the sample. This method is highly accurate and can identify species that are morphologically similar but genetically distinct.

DNA barcoding has several advantages over traditional identification methods. It allows for the identification of plankton at any life stage, including eggs and larvae, which are often difficult to distinguish morphologically. It also enables the detection of rare or cryptic species that might be overlooked in visual surveys. Furthermore, DNA barcoding can be applied to bulk samples, allowing researchers to identify multiple species simultaneously. This has made it an invaluable tool for monitoring biodiversity, tracking changes in plankton communities, and assessing the health of aquatic ecosystems.

6.2   Metagenomics: Unraveling the Complexity of Plankton Communities

While DNA barcoding is effective for identifying individual species, it can be limited when dealing with complex plankton communities where many species coexist. This is where metagenomics comes into play. Metagenomics involves sequencing the entire genetic material present in a sample, rather than targeting specific genes. This approach provides a comprehensive overview of the species composition and functional potential of a plankton community.

Metagenomics begins with the extraction of DNA from a bulk plankton sample, followed by high-throughput sequencing, which generates millions of DNA sequences. These sequences are then assembled and analyzed to identify the species present and their relative abundance.

Metagenomics can also reveal information about the metabolic capabilities of the community, such as nutrient cycling, photosynthesis, and the production of bioactive compounds.

One of the key benefits of metagenomics is its ability to capture the full diversity of a plankton community, including bacteria, archaea, viruses, and small eukaryotes, many of which are difficult to study using traditional methods. This holistic approach provides insights into the interactions between different species and their roles in the ecosystem. For example, metagenomics has been used to study the dynamics of harmful algal blooms, which can have devastating effects on marine life and human health. By analyzing the genetic composition of these blooms, researchers can identify the factors that trigger their formation and develop strategies for early detection and mitigation.

6.3 Environmental DNA (eDNA) Analysis: A Non-Invasive Window into Plankton Diversity

Another groundbreaking molecular technique for plankton identification is environmental DNA (eDNA) analysis. eDNA refers to genetic material that organisms shed into their environment through skin cells, mucus, feces, or other biological materials. By collecting and analyzing eDNA from water samples, scientists can detect the presence of plankton species without needing to capture or visually identify them.

eDNA analysis involves filtering water samples to collect the genetic material, extracting the DNA, and then amplifying and sequencing specific markers to identify the species present. This method is highly sensitive and can detect even low-abundance species, making it particularly useful for monitoring rare or elusive plankton. It also allows for non-invasive sampling, which is less disruptive to the ecosystem compared to traditional plankton nets.

eDNA analysis has proven to be a powerful tool for studying plankton diversity and distribution. It is being used to monitor invasive species, track changes in plankton communities due to climate change, and assess the impact of human activities such as pollution and overfishing. Moreover, eDNA analysis can be combined with other molecular techniques, such as DNA barcoding and metagenomics, to provide a more comprehensive understanding of plankton ecosystems.

6.4   The Future of Plankton Research: A Molecular Revolution?

The integration of molecular techniques into plankton research is revolutionizing our understanding of these vital organisms. DNA barcoding, metagenomics, and eDNA analysis are enabling scientists to identify plankton with greater precision, uncover hidden diversity, and explore the complex interactions within plankton communities. These techniques are also providing new tools for monitoring the health of aquatic ecosystems, predicting the impacts of environmental change, and informing conservation efforts.

As technology continues to advance, the potential applications of molecular techniques in plankton research are vast. Emerging methods such as single-cell genomics, which involves sequencing the DNA of individual plankton cells, promise to provide even deeper insights into plankton biology and evolution. Additionally, the development of portable sequencing devices and automated sampling systems could make molecular plankton identification more accessible and widespread, bringing these powerful tools out of the laboratory and into the field.

In conclusion, the use of molecular techniques for identifying plankton represents a major leap forward in our ability to study and understand the microscopic world. However, (and it is a very important however) we should keep in mind that taxonomist experts are still required and, in fact, are essential in the process of incorporating molecular tools into the nowadays research. What would be the point of isolating a sequence of an organism if afterward, nobody is letting you know what species is? Unfortunately, experts on taxonomy are fewer and fewer every day, and we may end up losing this extremely valuable discipline.

7

Resources and Further Reading

7.1 Recommended Books

Books are invaluable resources that provide comprehensive information on plankton biology, ecology, and identification. Here, you have a list of some I find useful (most free of charge).

"Plankton: A Guide to Their Ecology and Monitoring for Water Quality" by Iain Suthers and David Rissik (Eds). CSIRO Publishing. This book offers practical guidance on plankton ecology and techniques for monitoring water quality, making it a useful resource for both beginners and experienced researchers.

"A Guide to the Marine Plankton of Southern California" by David W. Folger. Free guide for educational purposes. Focused on marine plankton, this guide provides detailed descriptions and illustrations of plankton species found in Southern California waters.

"Freshwater Algae: Identification, Enumeration and Use as Bioindicators" by Edward G. Bellinger and David C. Sigee. Wiley-Blackwell Books. This book focuses on freshwater phytoplankton, offering detailed identification guides and discussing their use as bioindicators of water quality.

© The Author(s), under exclusive license to Springer Nature Switzerland AG 2024
A. Calbet, *The Amateur Plankton Researcher's Practical Guide*,
https://doi.org/10.1007/978-3-031-80248-5_7

"**Zooplankton of the Atlantic and Gulf Coasts: A Guide to Their Identification and Ecology**" **by William S. Johnson and Dennis M. Allen. The Johns Hopkins University Press.** A comprehensive guide to the identification and ecology of zooplankton found along the Atlantic and Gulf coasts of the United States.

"**The Wonders of Marine Plankton**" **by Albert Calbet.** This is my first outreach book in English and covers the basic concepts and peculiarities of the plankton, their ecology, and major roles in the ocean. I highly recommend this book because it is written in a quite comprehensible and engaging language.

"**Plankton in a Changing World: The Impact of Global Change on Marine Ecosystems**" **by Albert Calbet.** This is my second outreach book about plankton. It unravels the fundamental concepts of plankton ecology and their relevance while exploring the profound effects of global environmental changes on these vital organisms. I also highly recommend this one.

For children, I have: "**The Adventures of Pepo the Copepod**" **by Albert Calbet,** which describes the life of a copepod in a funny and comprehensible way, with characters and drawings. Ed. Amazon Kindle.

7.2   Online Databases and Identification Tools

Several online resources provide valuable tools for identifying plankton and accessing scientific information.

Plankton Portal A citizen science project where you can help classify images of plankton and learn about different species. https://www.planktonportal.org/.

AlgaeBase A comprehensive database of information on algae, including phytoplankton species. It includes taxonomy, distribution, and bibliographic references. https://www.algaebase.org.

Marine Planktonic Copepods A quite professional classification Web from a well-known zooplankton expert. https://copepodes.obs-banyuls.fr/en/.

7.3 Local and Online Communities for Plankton Enthusiasts

Connecting with communities of plankton enthusiasts can enhance your learning experience and provide opportunities for collaboration.

Phytoplankton Monitoring Network (PMN) A network of volunteers and professionals who monitor coastal phytoplankton and water quality. PMN

Citizen Science Projects Projects like **Plankton ID** on the **Zooniverse** platform allow you to participate in scientific research by helping to identify plankton in images. https://www.zooniverse.org.

Social Media Groups and Forums Platforms like Facebook, Instagram, TikTok, X, and YouTube host groups where plankton enthusiasts share findings, ask questions, and discuss topics related to plankton research. Search for groups such as "Amateur Microscopy" on Facebook, which share very beautiful plankton pictures. I manage a Facebook page: www.facebook.com/ZooplanktonEcology, where you can find information about plankton and ask me if you need help. Moreover, our group manages a YouTube Channel called Marine Zooplankton Ecology Lab.

Local Universities and Research Institutions Many universities and marine research institutions offer public outreach programs, workshops, and seminars on plankton and marine ecology. Contact local institutions to learn about opportunities to engage with the scientific community. I recommend the outreach materials of the Marine Sciences Institute, CSIC. www.icm.csic.es.

Professional Organizations Organizations like the **Association for the Sciences of Limnology and Oceanography (ASLO)** and the **International Society for the Study of Harmful Algae (ISSHA)** provide resources, conferences, and networking opportunities for professionals and enthusiasts alike. https://www.aslo.org | https://issha.org.

By leveraging these resources, you can expand your knowledge, improve your research skills, and connect with a broader community of plankton researchers and enthusiasts.

8

Practical Plankton Image Guide

About This Section

In this part of the guide, I provide images of the major groups of plankton. You will notice that I have deliberately avoided using species names, and for a simple reason: plankton taxonomy is highly complex, and the chances of misidentifying a species are quite high. Identifying species within each group, or even certain genera, can take years of dedicated study. Therefore, I have chosen to keep it simple by focusing on the major, most common groups without delving into overly complicated details. This guide is intended for amateur scientists, not professionals, so being able to distinguish a copepod from a rotifer or a dinoflagellate from a diatom is certainly an accomplishment. For some important groups, such as ciliates or copepods, I provide additional details about their classification. At the end of each major group, I also give insights into key physiological processes characteristic of that group.

8.1   Major Groups of Phytoplankton

Diatoms (Bacillariophyta) are single-celled algae encased in silica-based cell walls, forming intricate, glass-like structures. They thrive in both marine and freshwater environments, making them one of the most widespread types of phytoplankton. Diatoms are essential primary producers, contributing significantly to the carbon cycle in oceans, rivers, lakes, and streams. Their abundance makes them a critical component of aquatic ecosystems, especially in nutrient-rich waters.

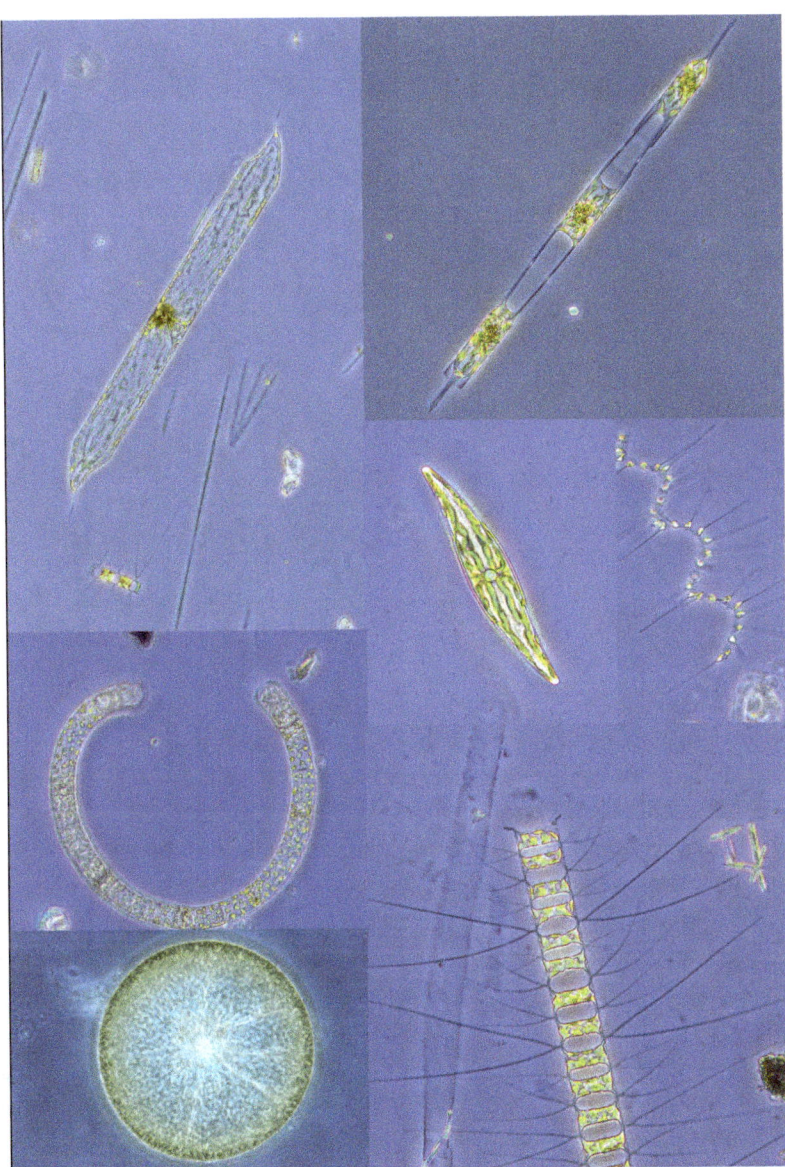

Marine diatoms

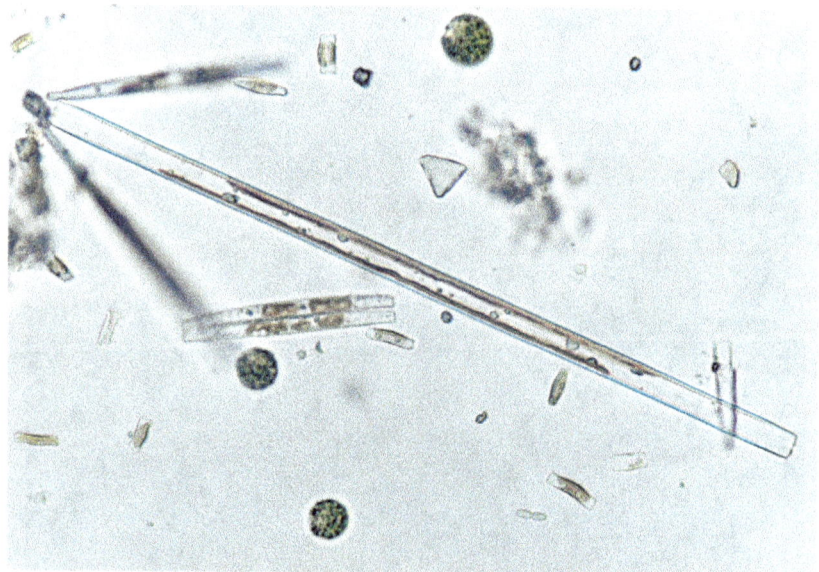

Freshwater diatoms

Dinoflagellates (Dinophyta) are single-celled organisms found mostly in marine environments, though some live in freshwater. They are equipped with two whip-like structures called flagella, which help them move. One flagellum wraps around the middle of the cell in a groove called the cingulum, allowing the organism to spin as it moves. The other flagellum extends backward from a groove called the sulcus, propelling the cell forward. Most dinoflagellates are covered by a tough outer layer called a theca, made of cellulose plates. This gives them their distinctive, often armored appearance. Some species lack these plates and have a more flexible, naked outer covering. Certain dinoflagellates are known because of their bioluminescence, while others can cause harmful algal blooms, commonly known as red tides. Predominantly found in marine environments, dinoflagellates can act as both primary producers and consumers. Their ability to form blooms can have both positive and negative impacts on marine ecosystems.

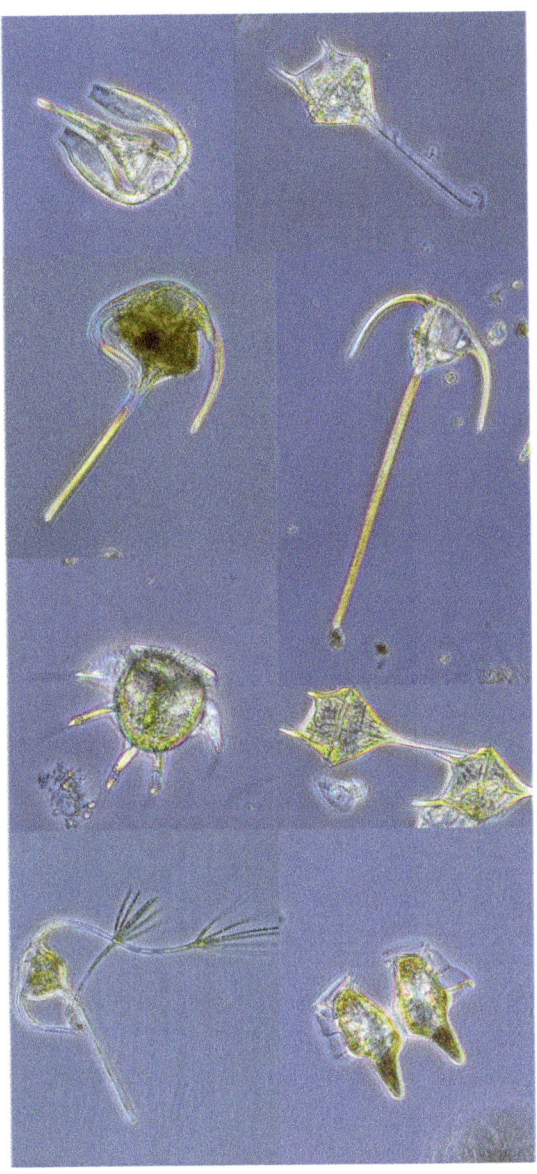

Marine dinoflagellates

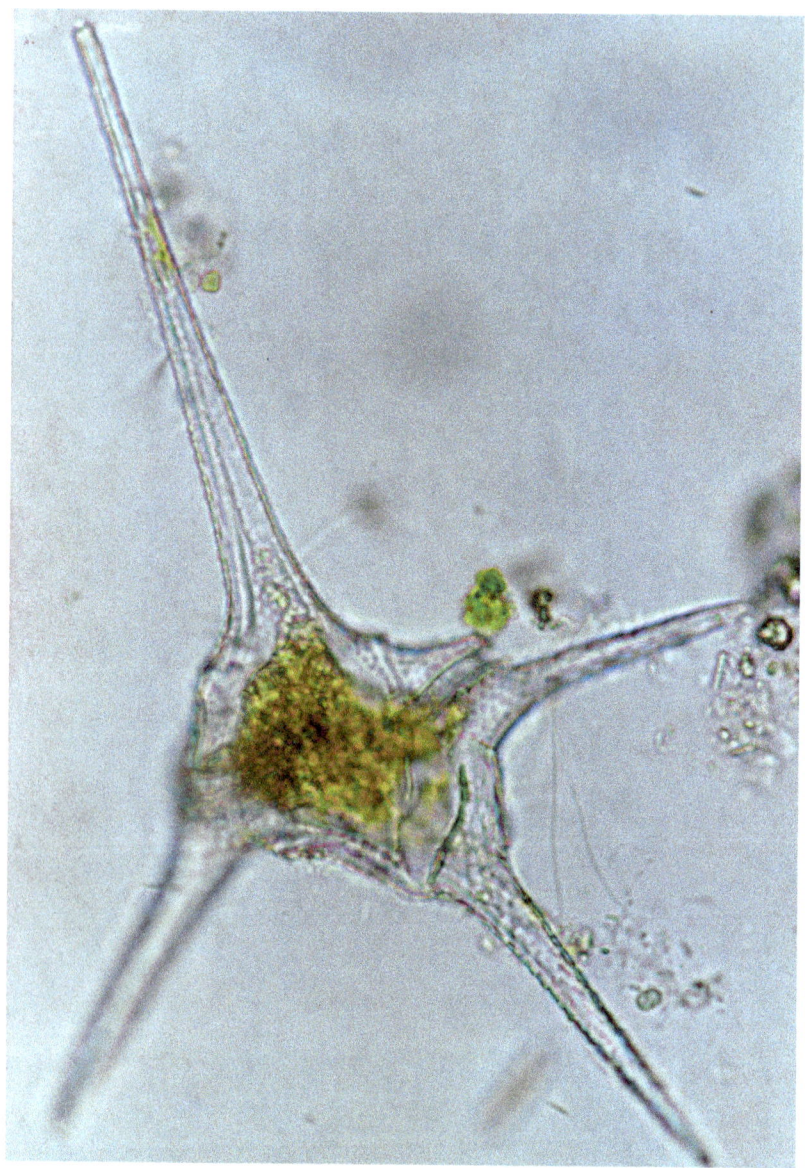

Freshwater dinoflagellate

Coccolithophores (Haptophyta) are single-celled algae covered in calcium carbonate plates, known as coccoliths. These organisms are mostly found in marine environments and are pivotal in the formation of chalk and limestone deposits. They play a vital role in the marine carbon cycle and can influence global climate by affecting carbon sequestration.

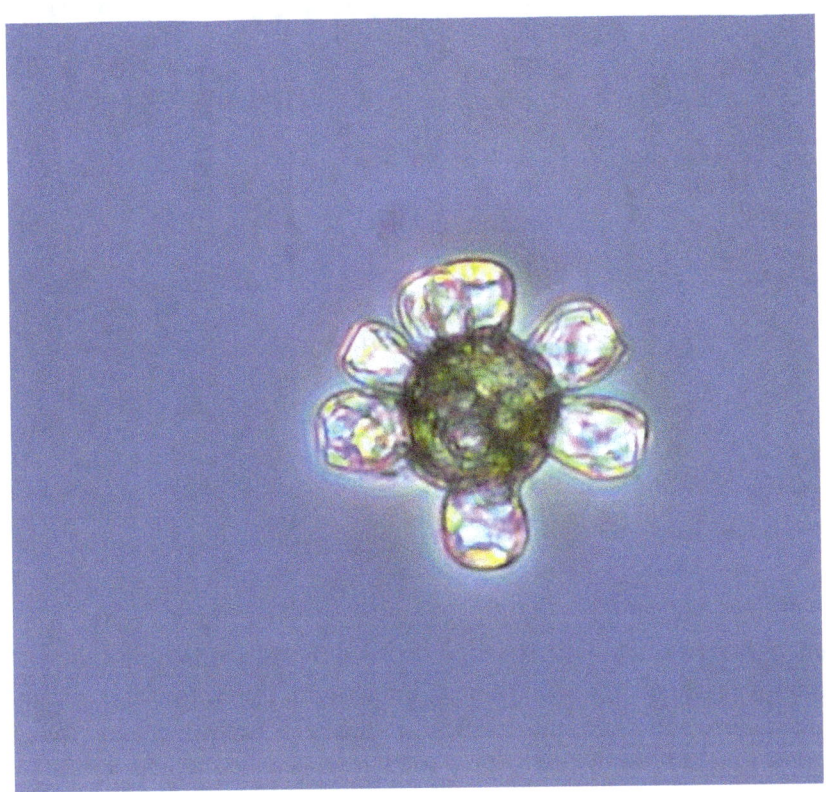

Marine coccolithophore

Cyanobacteria (Blue-green algae) are ancient, photosynthetic bacteria found in both marine and freshwater environments. Although they are bacteria, their ability to photosynthesize makes them functionally similar to algae. Cyanobacteria have been instrumental in shaping Earth's atmosphere by producing oxygen and are crucial for nitrogen fixation, converting atmospheric nitrogen into forms usable by other organisms. However, their blooms can sometimes be harmful, particularly in nutrient-rich waters.

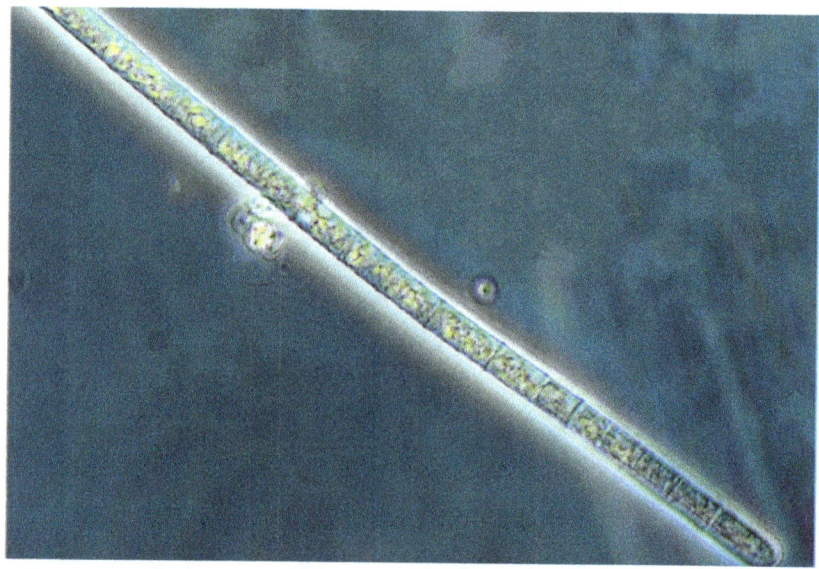

Marine cyanobacteria

Green Algae (Chlorophyta) encompass a diverse group of organisms, ranging from single-celled to multicellular forms, commonly found in freshwater lakes, rivers, and ponds, though some species are also marine. Green algae are important primary producers in freshwater ecosystems, supporting food webs and helping to maintain ecological balance.

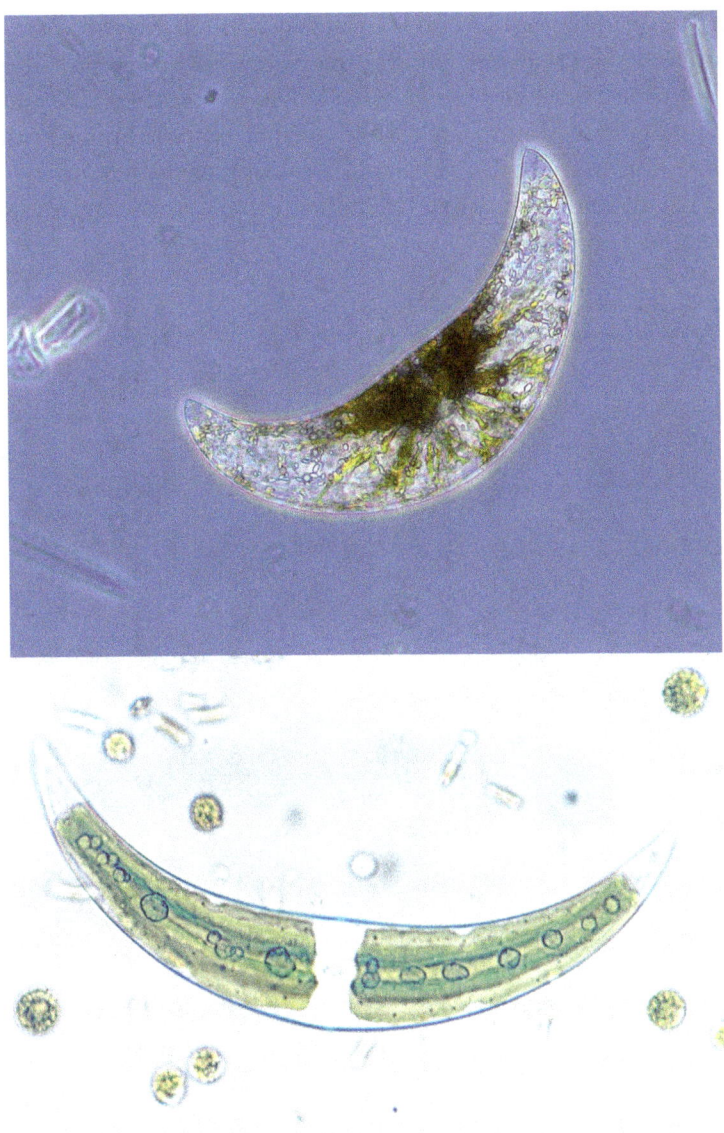

Marine (above) and freshwater (below) Chlorophyta of the same genus

Euglenoids (Euglenophyta) are flagellated, often green-colored algae that thrive in nutrient-rich, mostly freshwater environments. They are known for their flexibility in switching between photosynthetic and heterotrophic modes of nutrition. These organisms contribute to primary production in freshwater systems and are adaptable to fluctuating environmental conditions, making them resilient contributors to aquatic food chains.

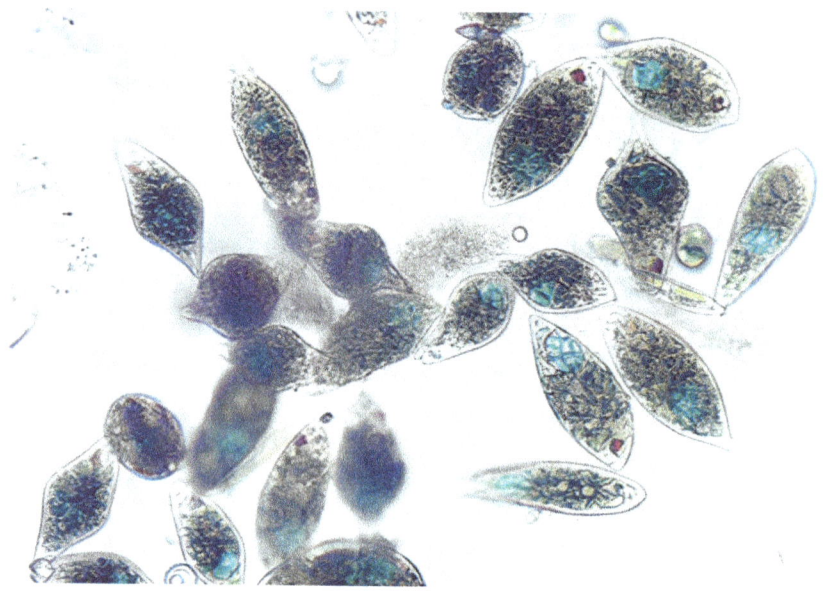

Freshwater euglenoids

8.2 Major Groups of Small Zooplankton (Protozoans and Small Metazoans)

Small zooplankton, i.e., these organisms that you can differentiate only by using a microscope, are usually unicellular (protozoans) or tiny metazoans, such as rotifers and other small worms. Free-living protozoans are diverse single-celled organisms that thrive in various environments, including freshwater, marine habitats, and soil. These protozoans are not reliant on a host for survival, unlike parasitic protozoans. Here is a brief overview of the major groups:

Amoeboids are protozoans that move and feed by extending their cytoplasm to form pseudopodia (false feet). These organisms, like *Amoeba proteus*, are common in freshwater environments. They capture food particles through phagocytosis and can adapt to various conditions by forming cysts. This group can present naked or testate forms. Testate (or thecate) amoebae (often called shelled amoebae) are characterized by their protective outer shell, or "test." This shell can be made of various materials, including organic substances, minerals, or particles gathered from the environment. Testate amoebae are commonly found in freshwater, soils, and mosses, playing an important role in nutrient cycling and microbial food webs by feeding on bacteria, algae, and small particles. They extend pseudopodia (temporary projections) through openings in their shell to move and capture food.

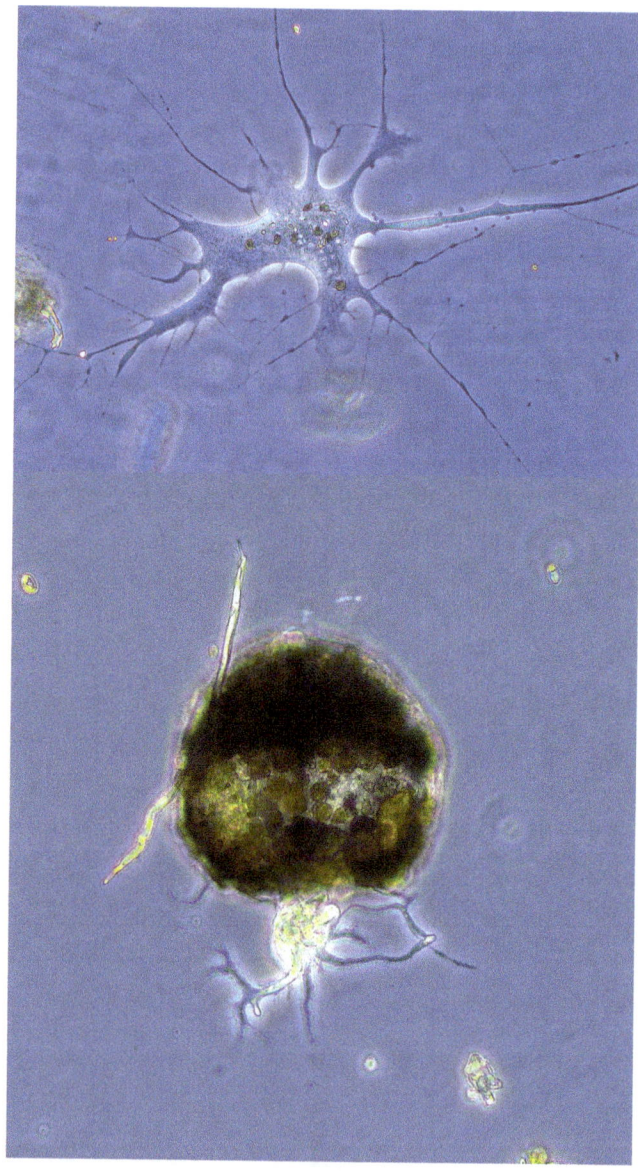

Marine naked (above) and testate (below) amoeba

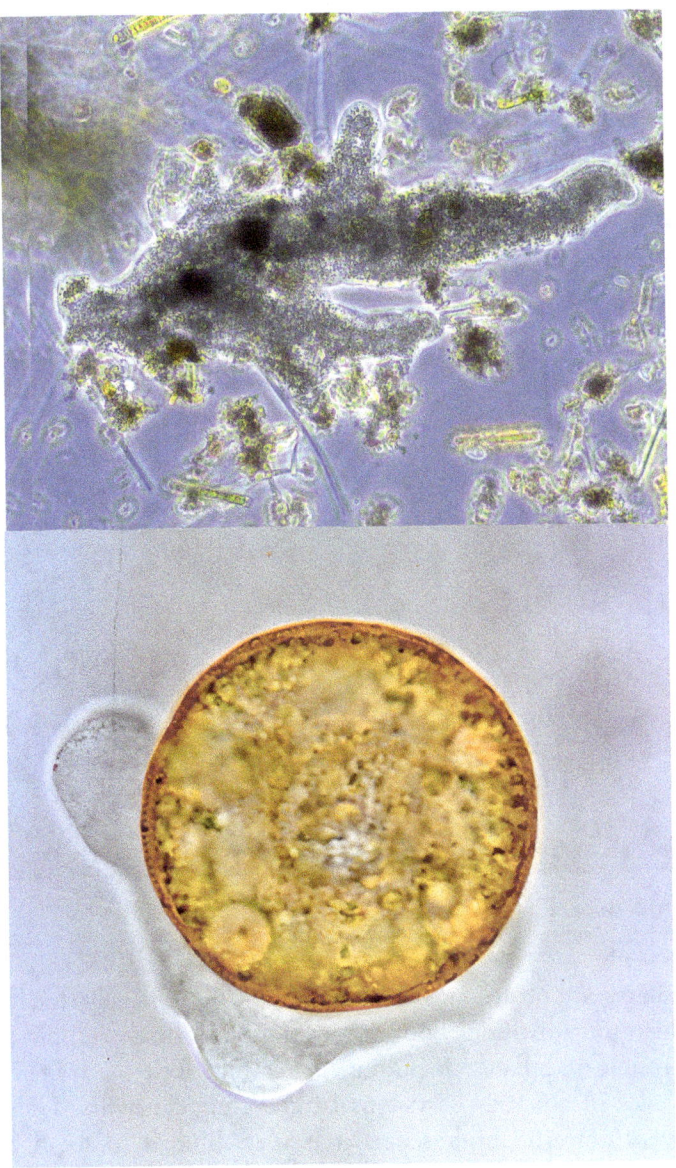

Freshwater naked (above) and testate (below) amoeba

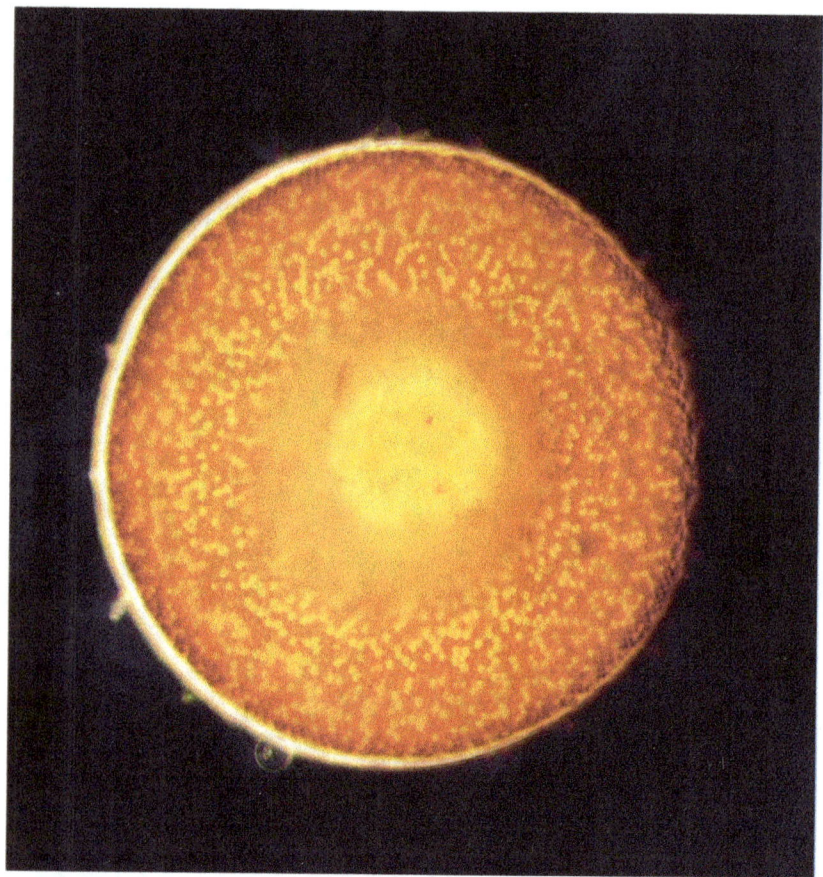

Detail of the shell of a testate amoeba

Flagellates are protozoans that use one or more flagella for locomotion. This group includes both autotrophic and heterotrophic species. Heterotrophic flagellates, including many that feed on bacteria, play critical roles in aquatic ecosystems by recycling nutrients. Unless you have a very powerful microscope, it is very difficult to see flagellates. Here, I include a picture of a flagellate community at 1000x magnification under confocal microscopy.

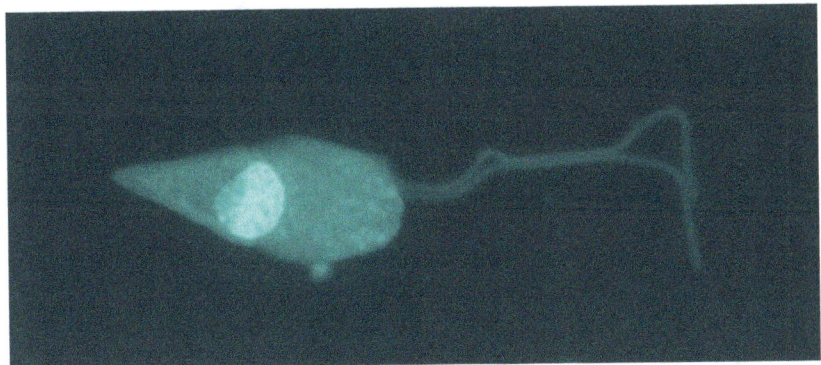

Image of a eukaryotic flagellate taken with confocal microscopy. Author Irene Forn

Ciliates are characterized by the presence of numerous cilia, used for movement and feeding. Ciliates, such as *Paramecium*, are highly diverse and complex, with specialized structures like an oral groove for feeding. They thrive in freshwater, marine environments, and soils.

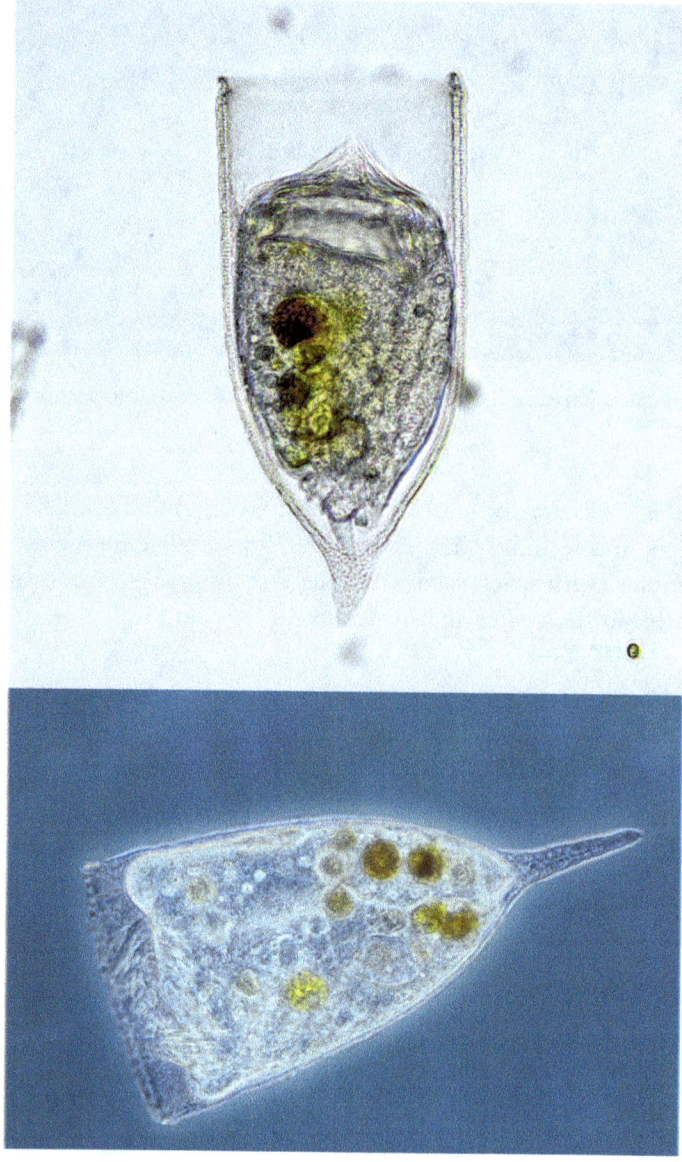

Tintinnid ciliates are characterized by possessing a lorica or shell that covers most of the body

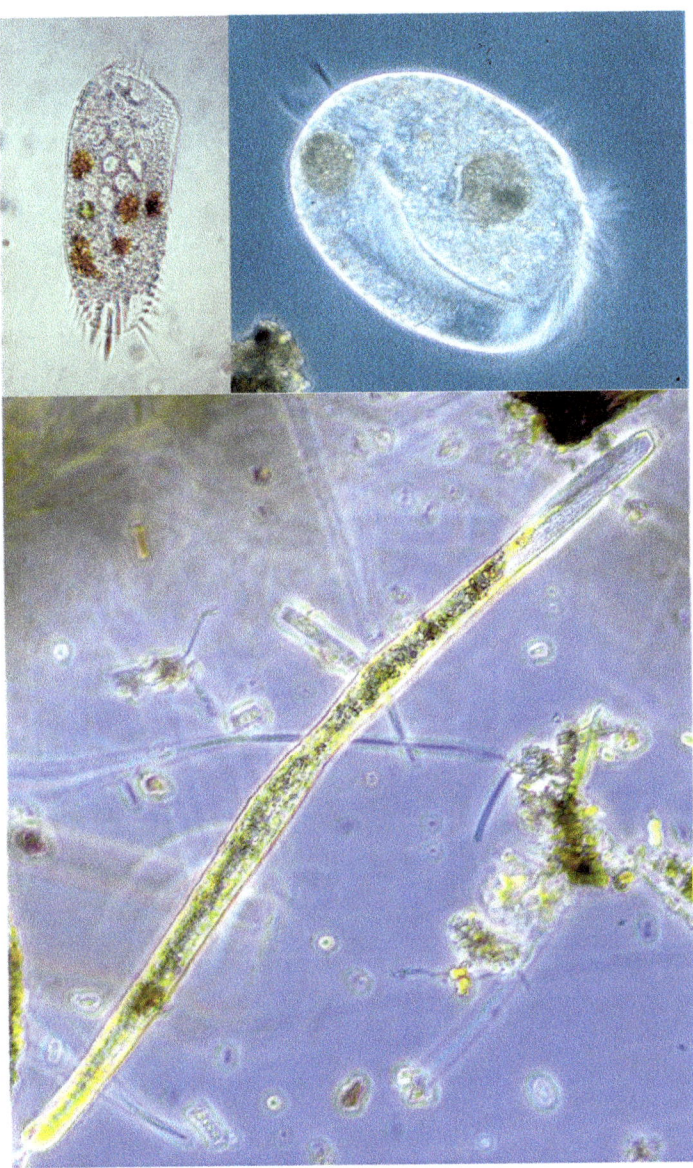

Freshwater ciliates

The Major Groups of Ciliates

Ciliates are classified into several major groups based on their morphological and genetic characteristics. Some of the prominent groups include:

- **Heterotrichea:** Known for their large size and often elaborate structures, members of this group, like *Stentor*, are among the most visually striking ciliates.

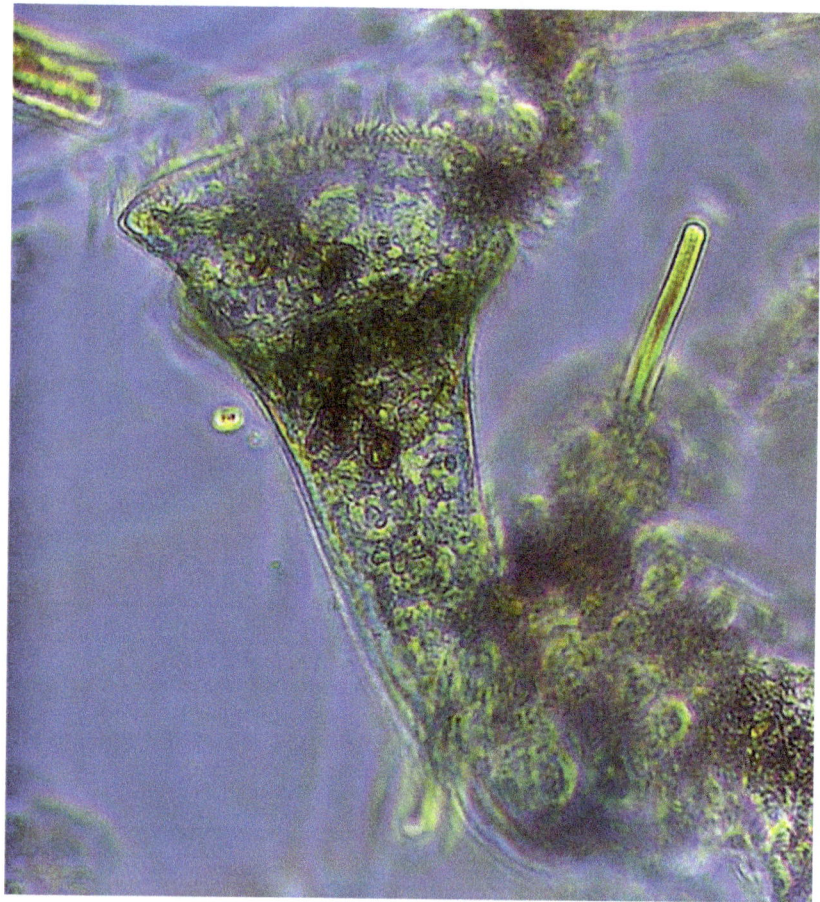

Heterotrichea ciliate

- **Oligohymenophorea:** This group includes well-known ciliates like *Paramecium* and *Vorticella*, which are often studied in biology labs for their straightforward yet fascinating behaviors.

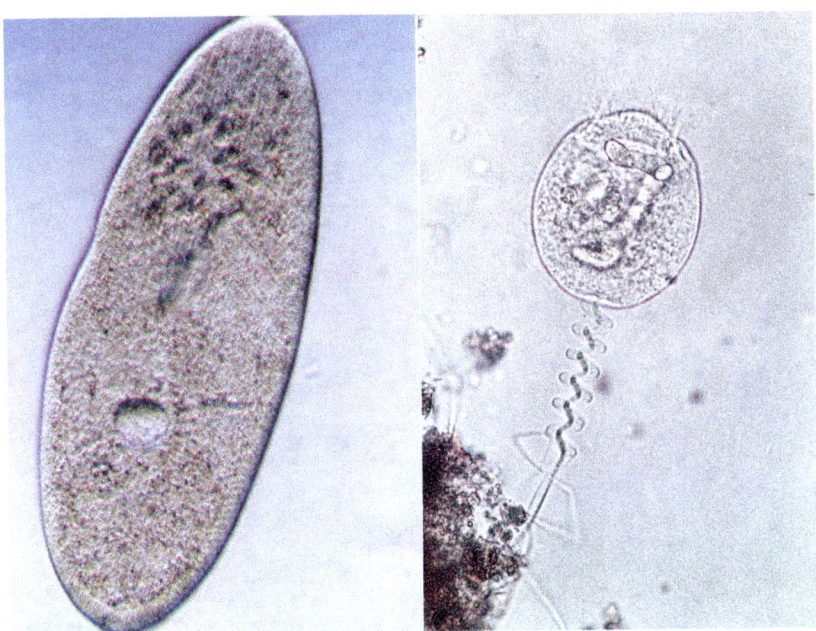

Oligohymenophorea ciliates

- **Spirotrichea:** Recognized for their complex ciliary structures, these ciliates often have bristles or cirri that help in movement and feeding.

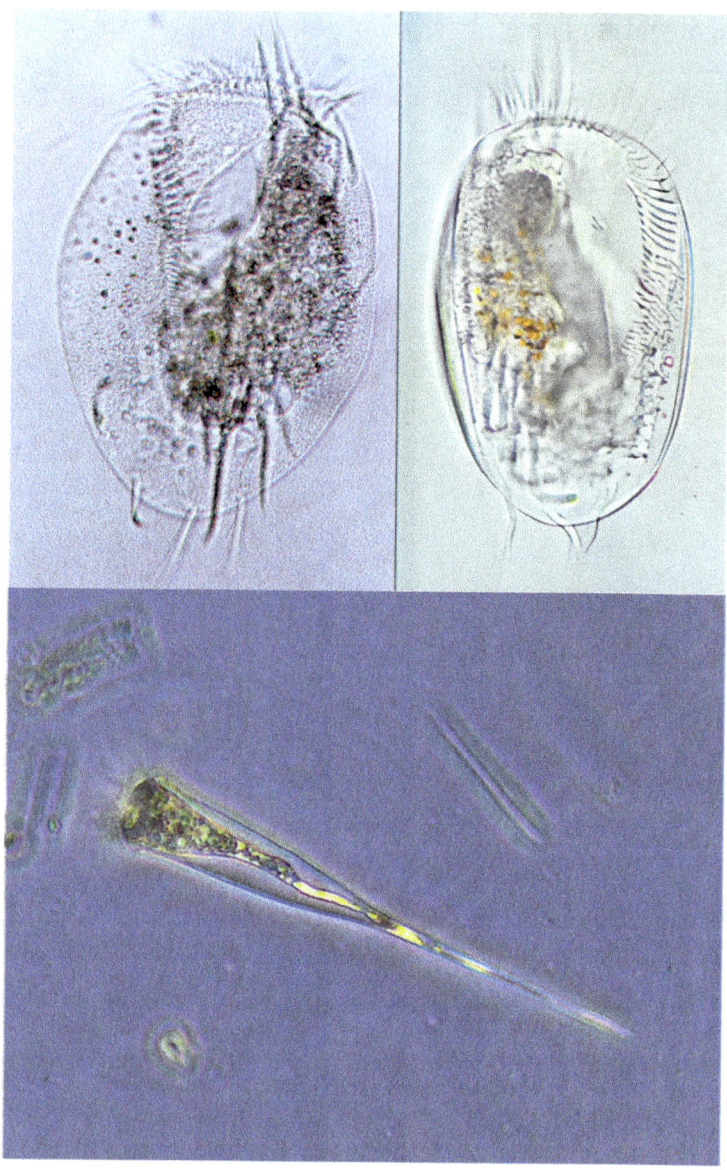

Spirotrichea ciliates

- **Litostomatea:** These ciliates are predominantly predatory, with some members, like *Didinium* or *Lacrymaria*, known for their voracious appetites.

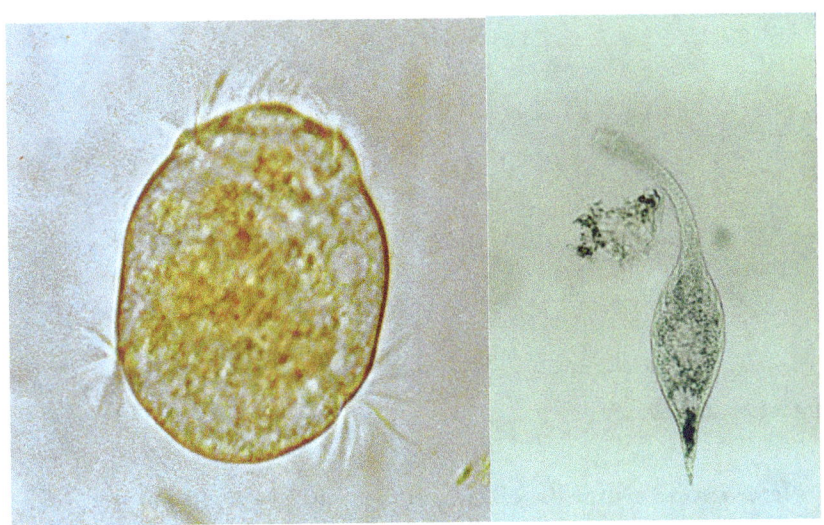

Litostomatea ciliates

- **Protostomatea.** They are called so because of their distinctive feeding structures, particularly the "protostome" (meaning "first mouth"). In these ciliates, the mouth-like opening (cytostome) is located near the front (anterior) of the cell, allowing them to capture and ingest food. The genus *Coleps* is characterized by its armored structure.

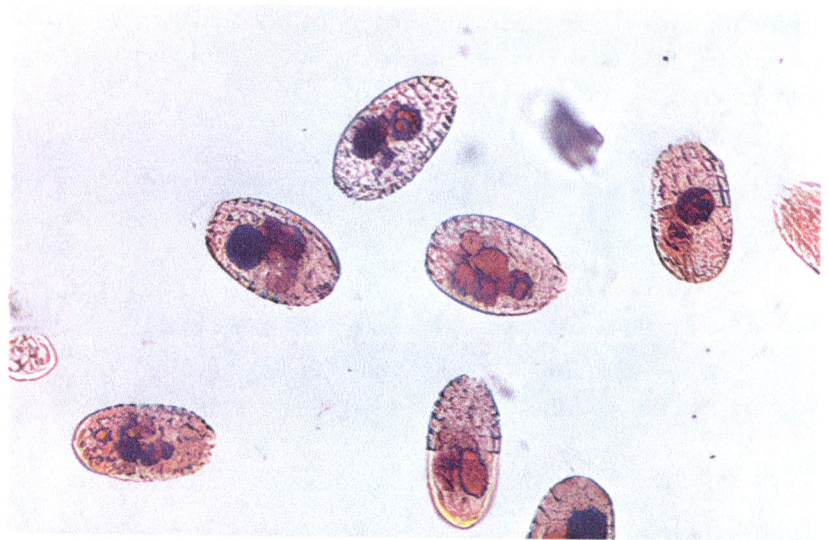

Protostomatea ciliates

Heterotrophic dinoflagellates are a group of flagellated protozoans. While many dinoflagellates are photosynthetic, heterotrophic species obtain nutrients by ingesting other organisms. They are common in marine environments and are key in planktonic food webs, particularly in nutrient-poor waters.

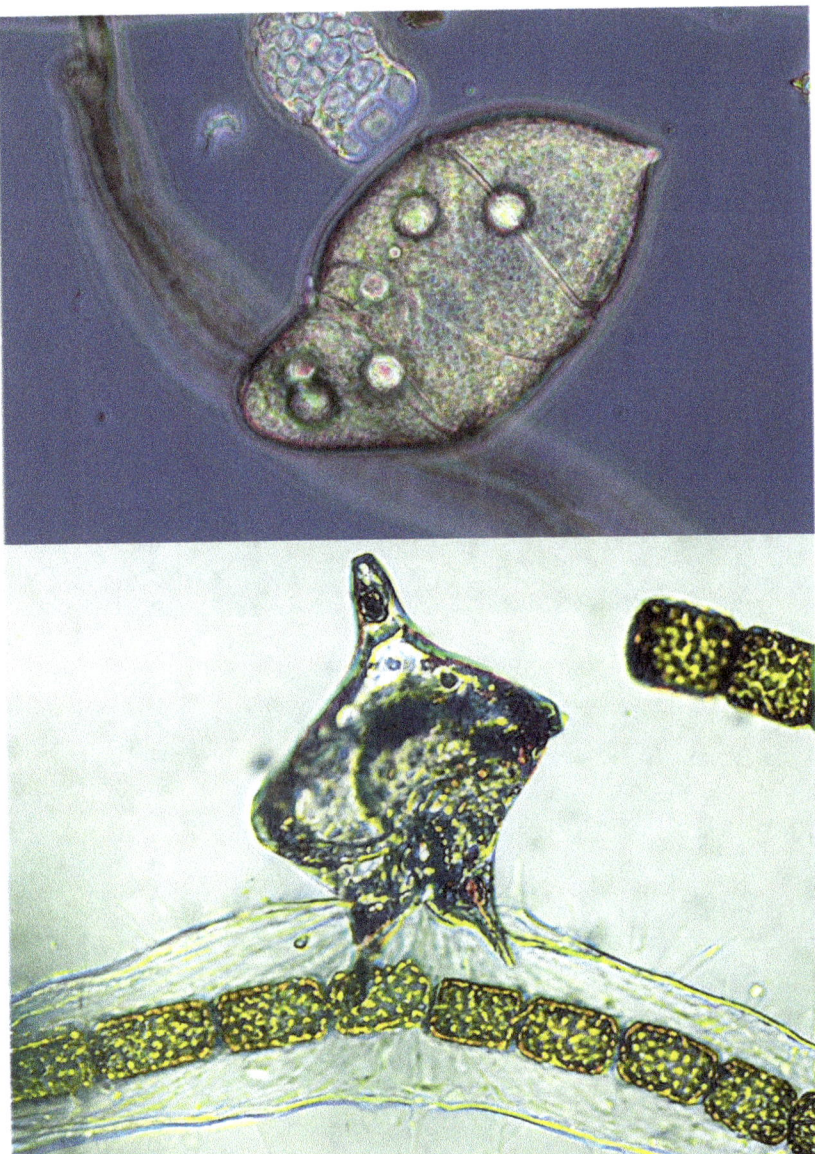

Above, heterotrophic dinoflagellates with vacuoles. Below, heterotrophic dino-flagellate feeding on a diatom chain

Radiolarians and Heliozoans: Radiolarians and Heliozoans are often grouped together because of their similar morphology, particularly their use of axopodia. However, radiolarians are typically marine with silica-based skeletons, while heliozoans are more often freshwater organisms without the silica skeletons. They capture food with axopodia, supported by microtubules.

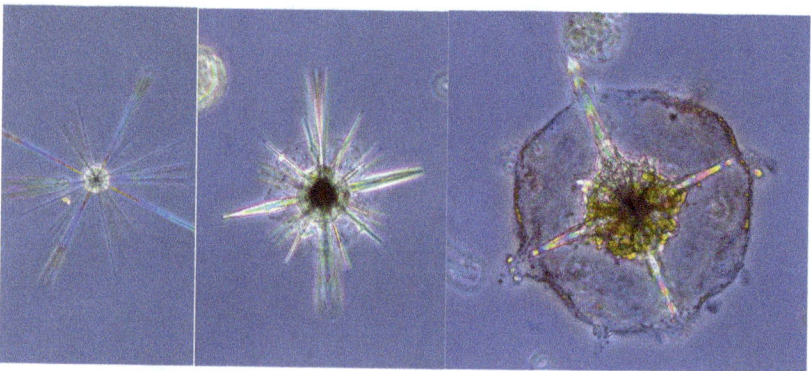

Marine radiolarians

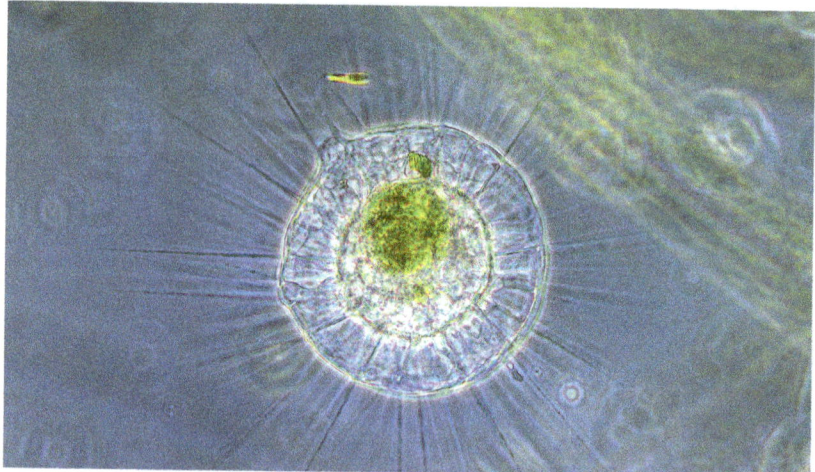

Freshwater heliozoan

Foraminiferans are amoeboid protozoans with calcium carbonate shells, known as tests. These primarily marine protozoans contribute to marine sediments, forming deposits like chalk and limestone. They extend pseudopodia through the openings in their tests to capture food.

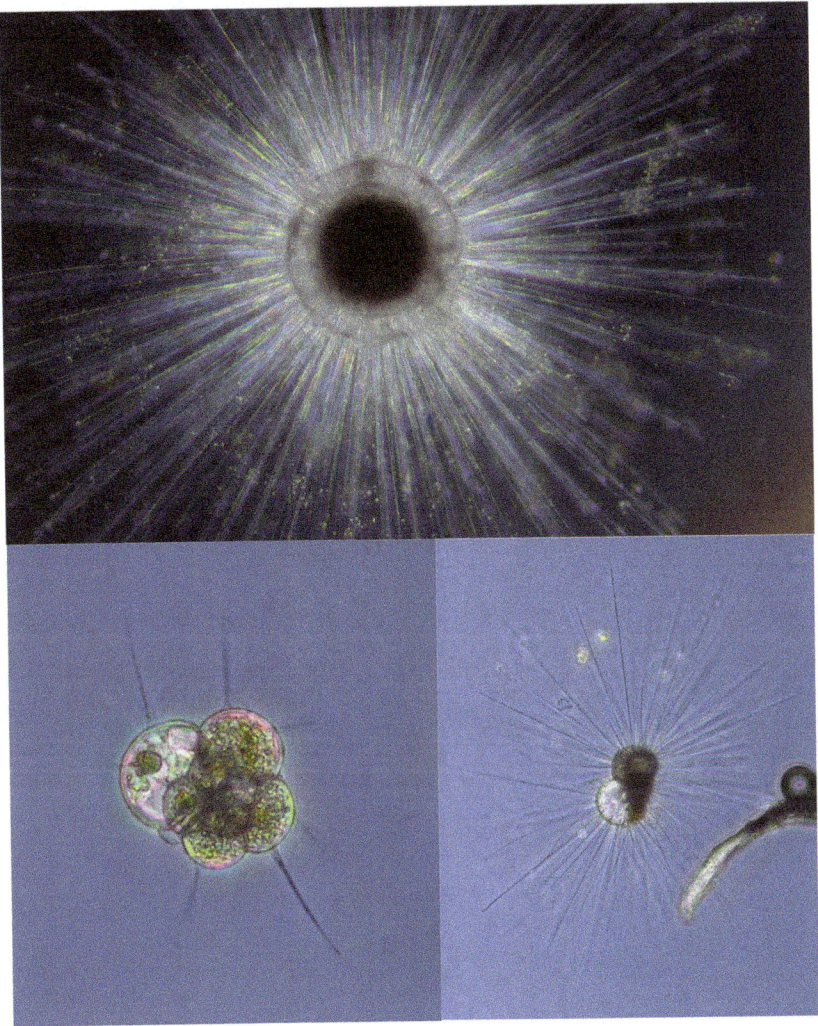

Foraminiferans

Protozoans' Unique Cellular Structure

One of the most intriguing aspects of ciliates and other protozoans is their **dual-nucleus system**, which sets them apart from most other eukaryotes. Ciliates possess two types of nuclei within each cell: a large macronucleus and a smaller micronucleus. The macronucleus is responsible for the day-to-day operations of the cell, such as metabolism and growth, while the micronucleus holds the genetic material that is passed on during reproduction.

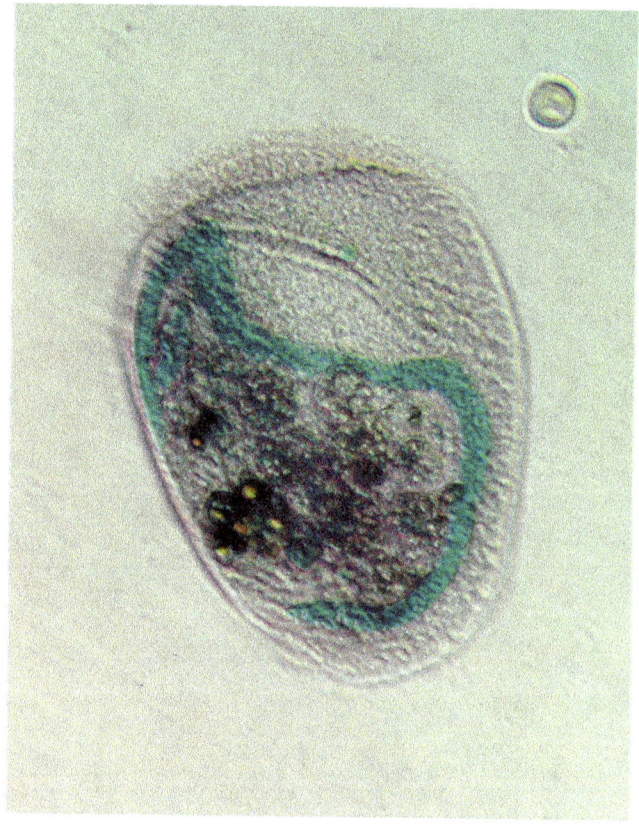

Ciliate, showing the macro and the micronucleus in green

The cell structure of protozoans is further complicated by specialized organelles. For instance, **food vacuoles** are formed when ciliates engulf their prey, typically bacteria or smaller protists. The food vacuoles travel through the cell, digesting the contents and absorbing nutrients.

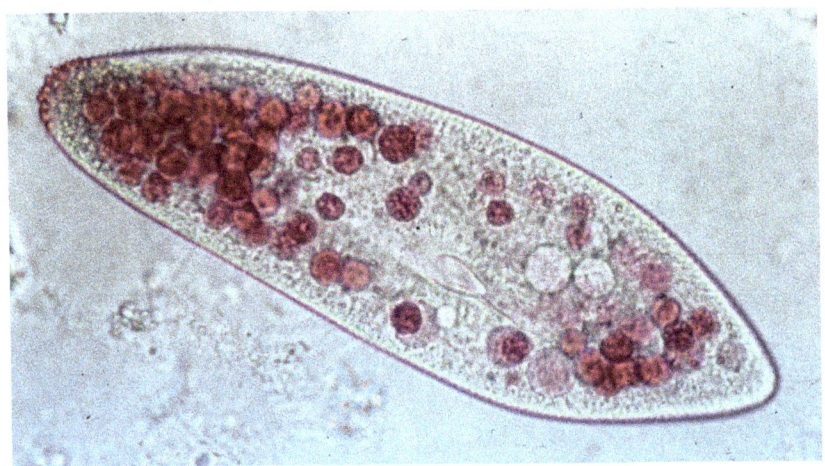

Food vacuoles in red

They also possess **contractile vacuoles** that help regulate water balance, a crucial function for their survival in varying aquatic conditions.

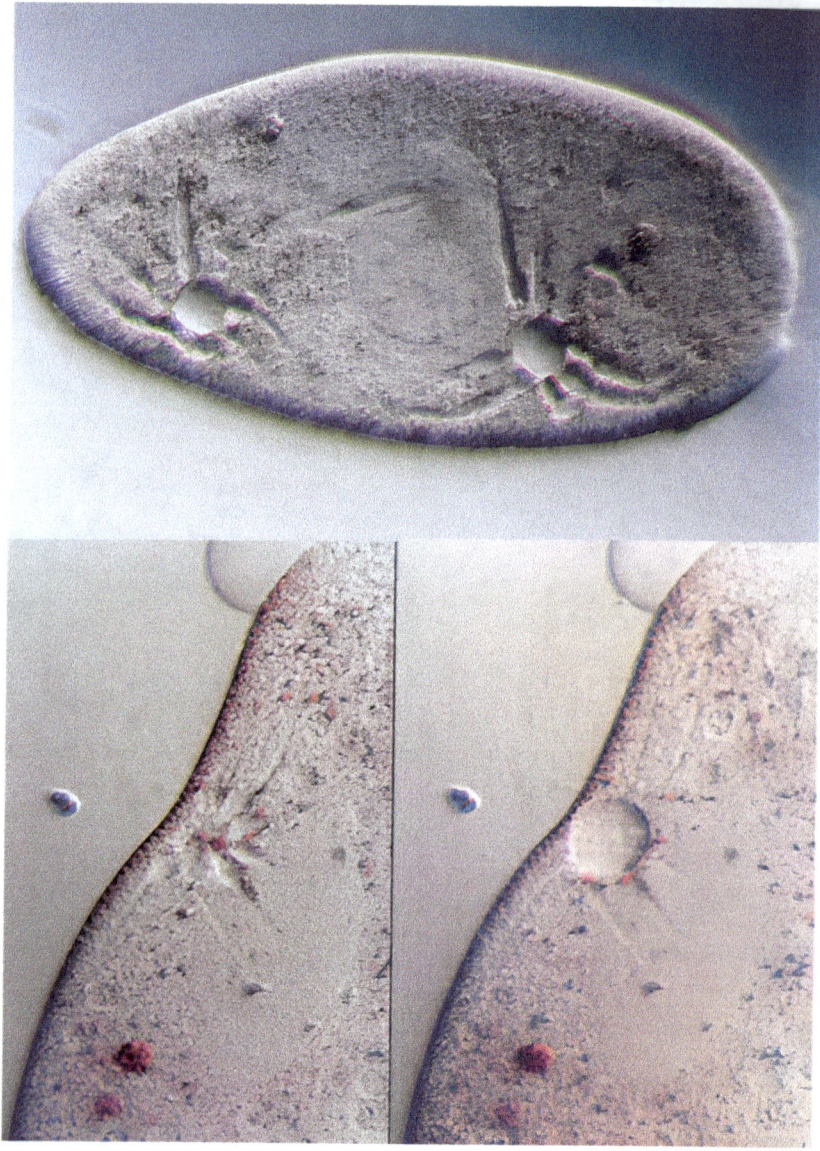

Contractile vacuoles. The image was obtained with a very inexpensive technique. It involves placing a coin or another small object over of the light source. By playing with the shadows, you can get 3D-looking images

Feeding Strategies

While many protozoans are heterotrophs, feeding on bacteria, algae, and other small organisms, they display a wide range of feeding strategies. Some protozoans are **mixotrophs**, combining the ingestion of food with photosynthesis, thanks to their symbiotic relationships with photosynthetic microbes. Others have even been observed feeding on viruses, showcasing their diverse dietary habits.

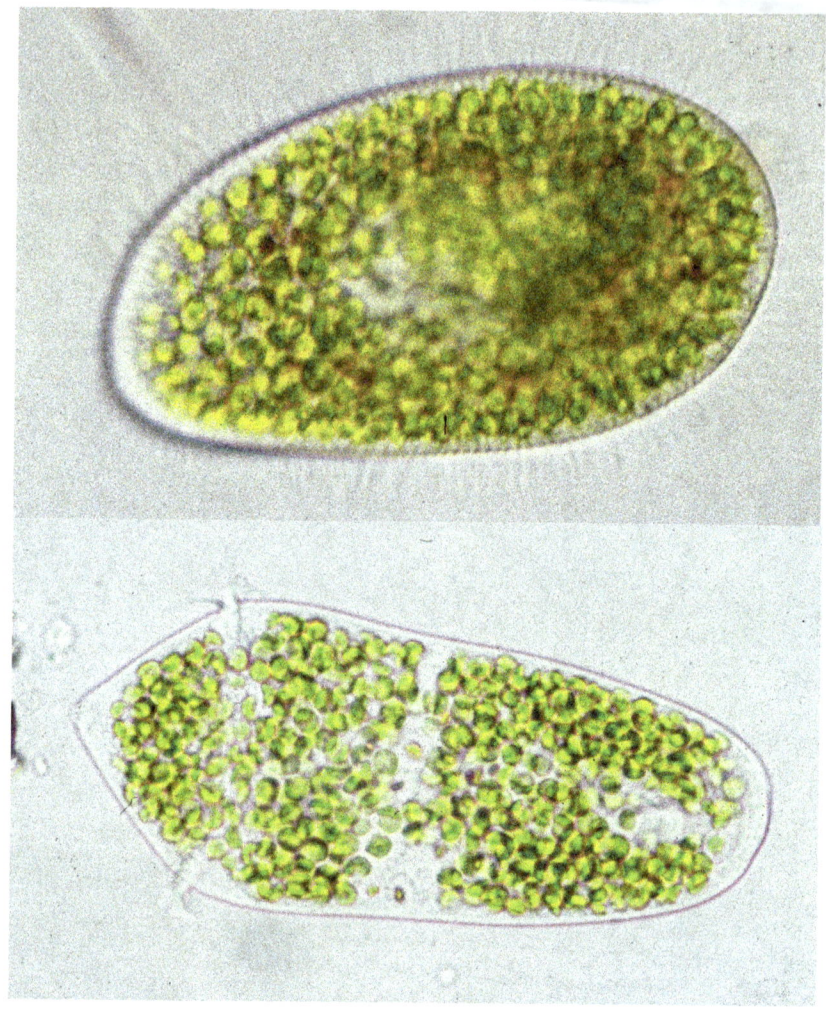

Mixotrophic ciliates

Certain protozoans are predators, preying on other protozoa, while some are parasites, though only one species, *Balantidium coli*, is known to cause disease in humans. The diversity in feeding strategies among ciliates highlights their role as key players in the aquatic food web.

Protists exhibit a remarkable diversity of feeding strategies, each adapted to their specific ecological niches. Among protozoans, some like **amoebas** use **engulfment** or phagocytosis, where they extend their membrane to encircle and ingest food particles, such as bacteria or other small prey. **Ciliates** like *Paramecium* employ cilia to sweep food into their oral groove, where it is enveloped in a food vacuole for digestion.

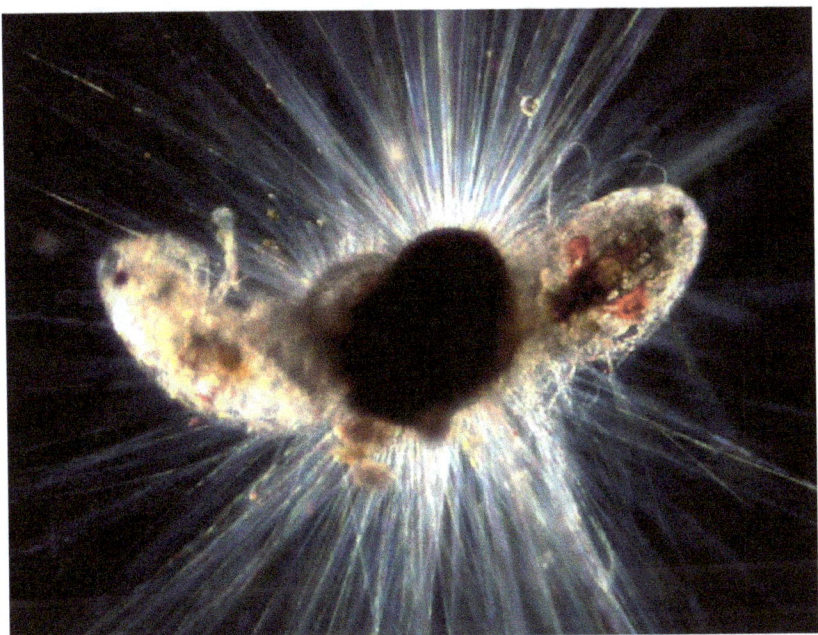

A foraminiferan feeding on two copepods

Dinoflagellates, a particularly versatile group, showcase a range of unique feeding techniques. By **engulfment**, some dinoflagellates capture and ingest entire prey cells, similar to the method used by amoebas. Others practice **tube feeding**, where they extend a tube-like structure to suck out the cell contents of their prey, allowing them to consume larger organisms without engulfing them. Another fascinating method is **pallium feeding**, where the dinoflagellate extends a sheet-like structure called the pallium over its prey, digesting it externally before absorbing the nutrients.

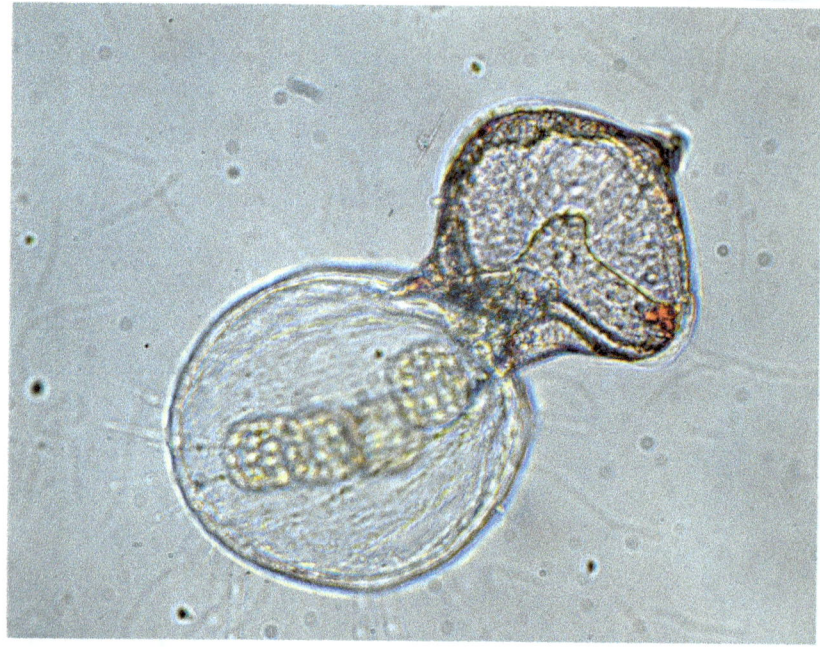

A heterotrophic dinoflagellate pallium-feeding on a diatom chain

These varied feeding strategies not only highlight the adaptability and survival mechanisms of protists but also underscore their critical role in aquatic ecosystems, where they contribute to nutrient cycling and energy transfer across the food web.

Reproduction and Sexual Phenomena

Protozoans typically reproduce asexually through a process known as binary fission, where the cell splits into two identical daughter cells.

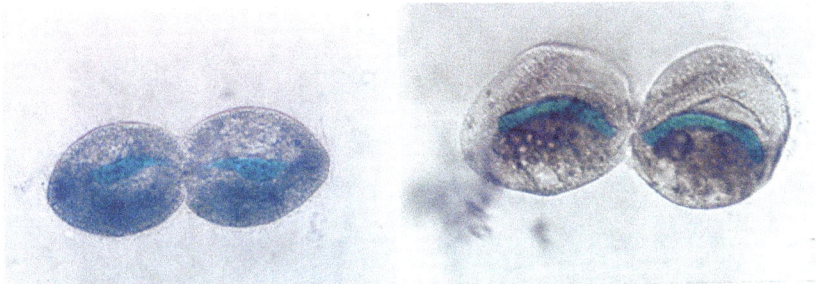

Binary fission in ciliates. The nucleus is stained in green

However, they also engage in a unique form of sexual reproduction called conjugation. During conjugation, two protozoans of compatible mating types exchange genetic material, resulting in genetic recombination and increased diversity within the population. This process does not increase the number of individuals but rejuvenates the cell line, helping to prevent aging.

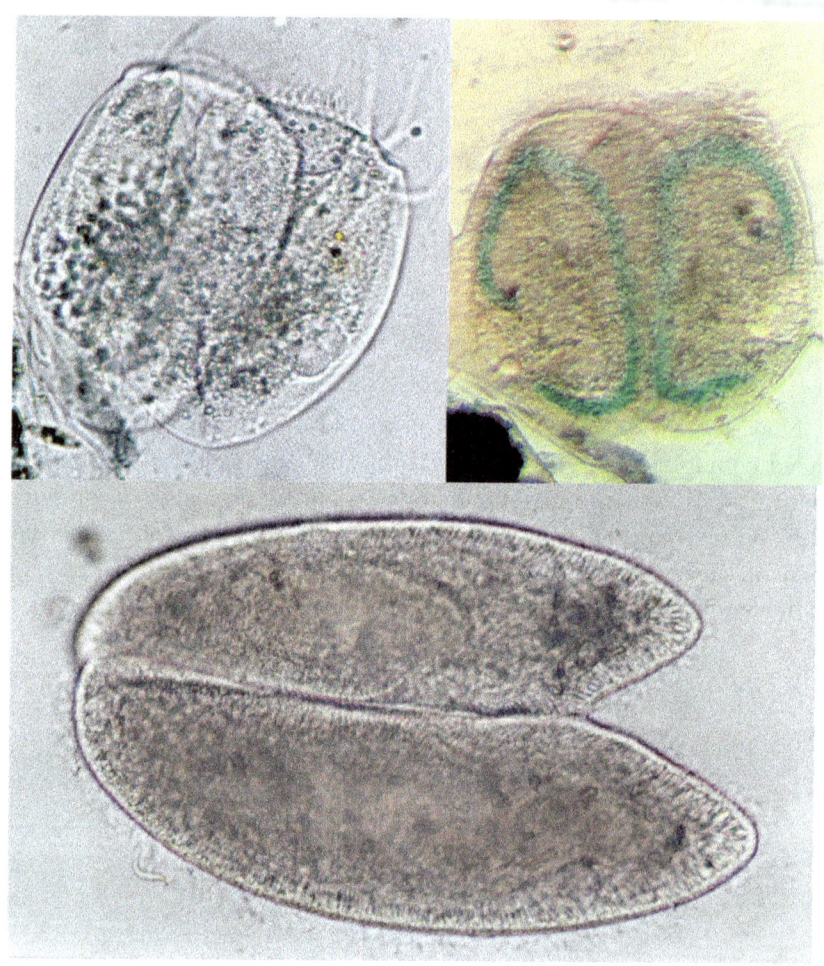

Conjugation in ciliates. Note the green nucleus in the image in the right panel, above

Rotifers and other worms are microscopic pluricellular organisms primarily found in freshwater, though some marine species exist. They have a unique ciliated structure called a corona used for feeding and loco-motion. Rotifers feed on small particles, like bacteria and phytoplankton, and are an essential food source for fish larvae and other micro-predators. Other common groups of worms commonly seen in water samples are nematodes, and polychaeta and their larva.

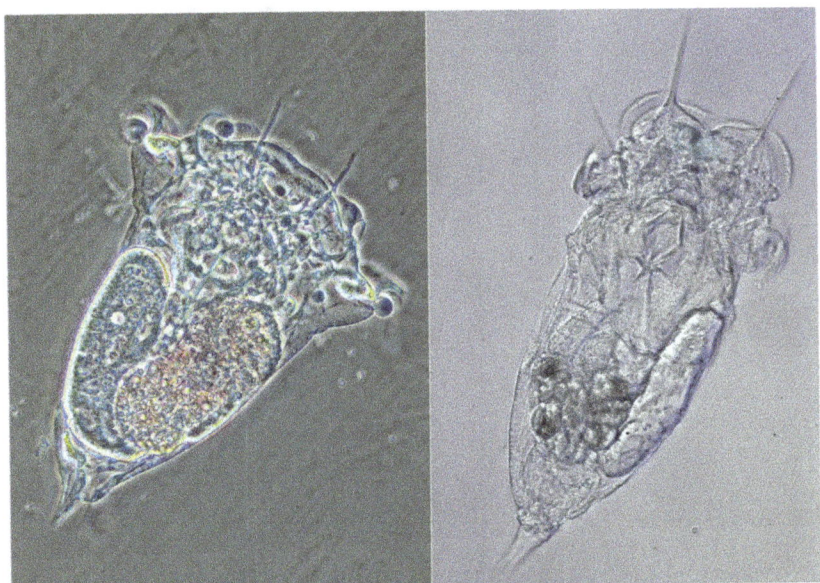

Marine rotifers

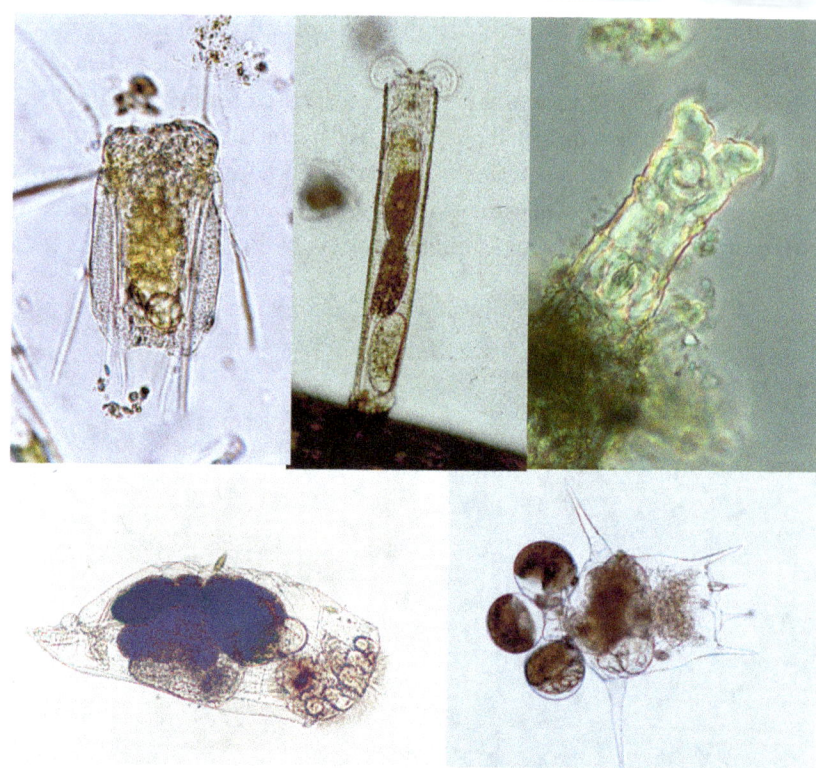

Freshwater rotifers

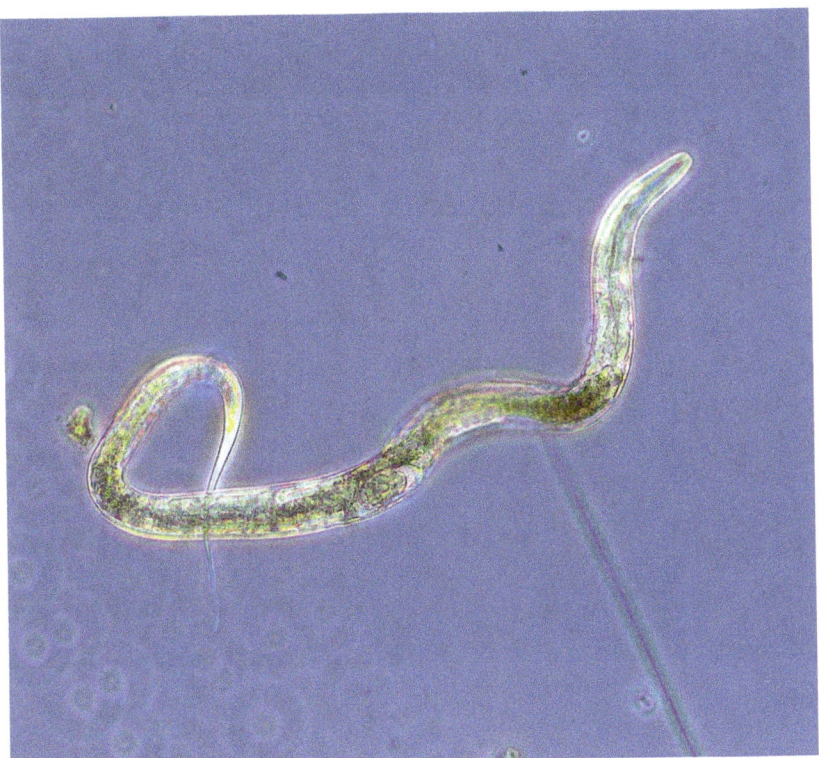

Nematoda

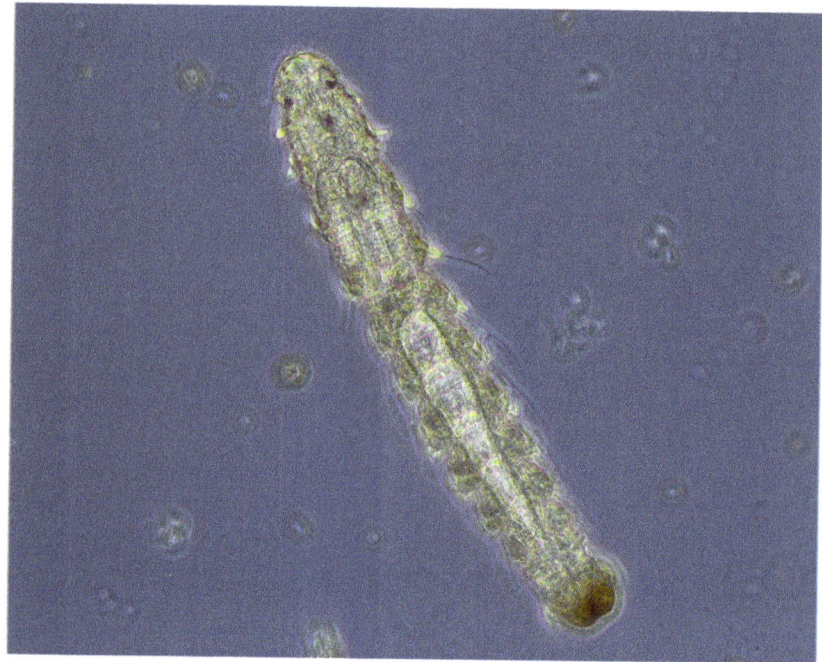

Polychaeta

8.3   Major Groups of Large Zooplankton (Metazoans)

Most of these organisms can be viewed under the stereomicroscope or even by the naked eye. They include crustaceans, jellyfish, tunicates, and some larvae of benthic organisms, among others.

Copepods are small, shrimp-like crustaceans that are among the most abundant animals on Earth, particularly in marine environments, but also prevalent in freshwater. As primary consumers, copepods feed on phytoplankton and microzooplankton, serving as a vital food source for fish, whales, and other marine creatures.

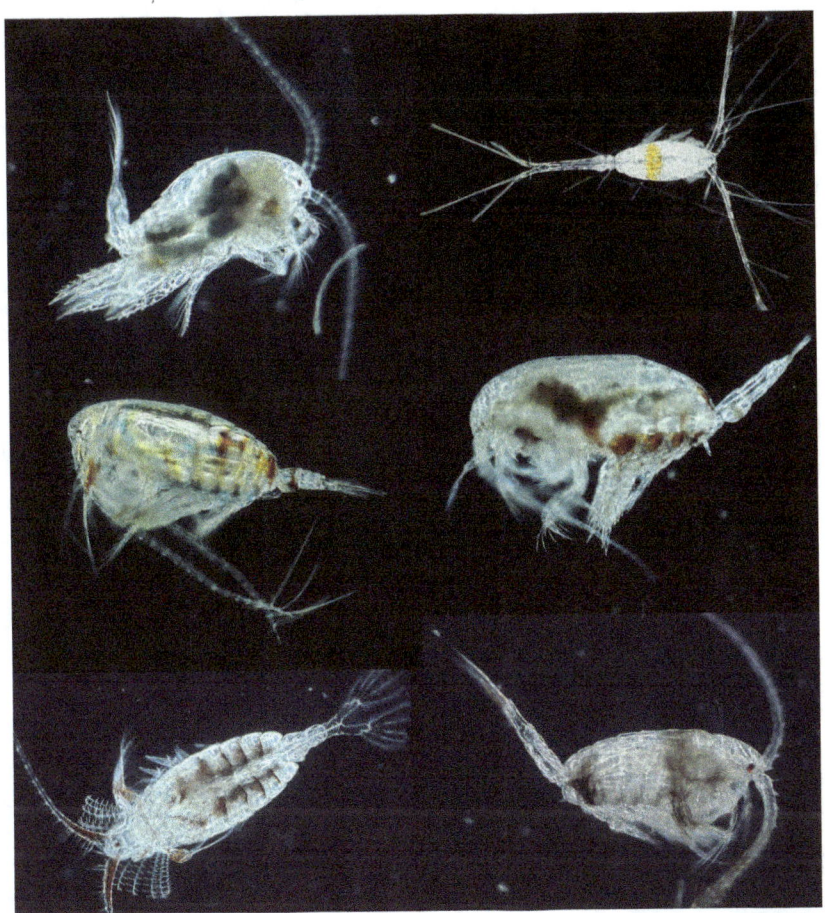

Marine copepods

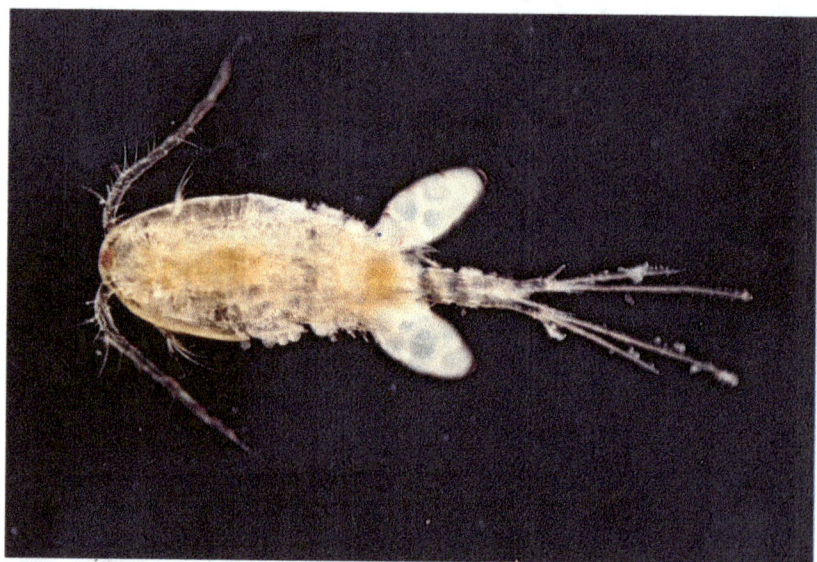

Freshwater copepod

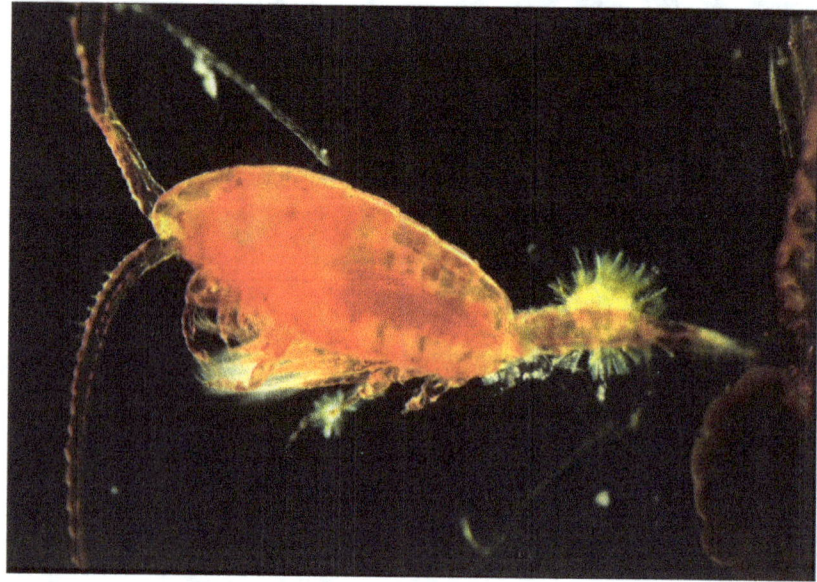

Hypersaline environments copepod

Copepod Major groups: Free-living copepods are traditionally divided into three major groups: **Calanoida**, **Cyclopoida**, and **Harpacticoida**. Calanoids are typically found in open waters and are key players in marine ecosystems, often serving as primary food for fish larvae. Cyclopoids are more versatile, inhabiting both freshwater and marine environments; they are important predators of small aquatic organisms. Harpacticoids are mostly benthic, living on the sea or lake floor, where they feed on detritus and microorganisms, playing a crucial role in nutrient cycling. Together, these groups support aquatic food webs and maintain ecosystem health.

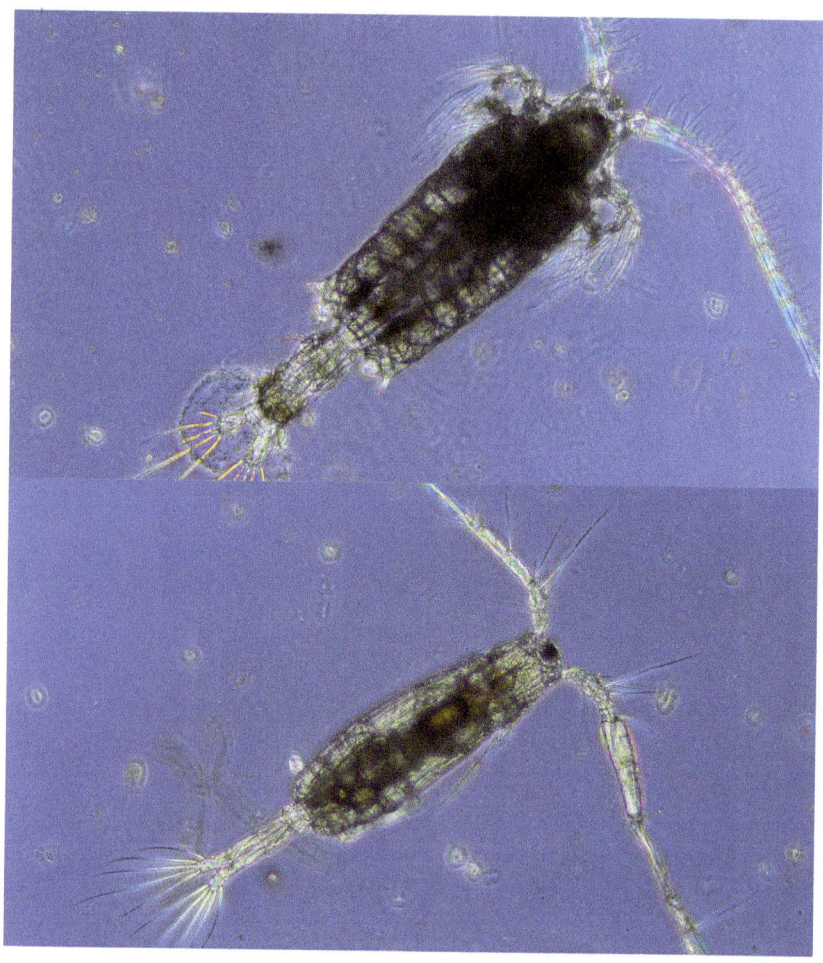

Marine calanoid copepods

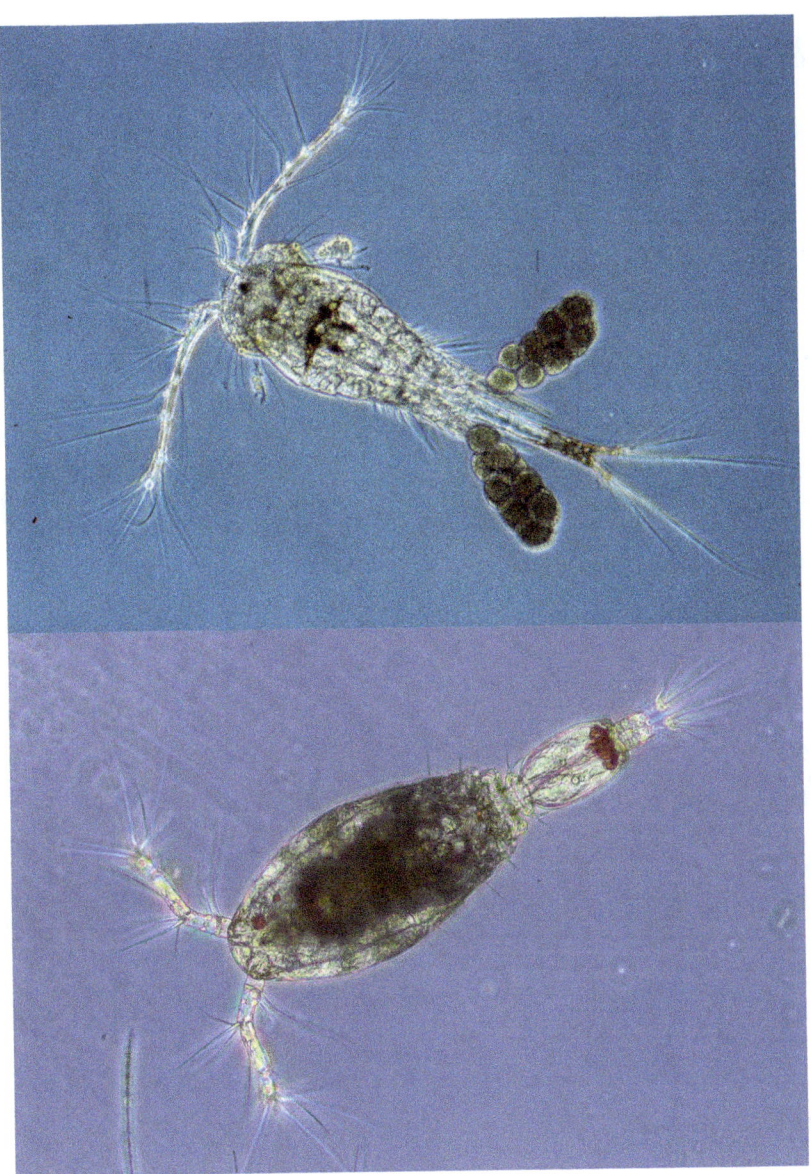

Marine cyclopoid copepods

Freshwater cyclopoid copepod

Marine harpacticoid

Freshwater harpacticoid

Copepods' Life Cycle: Copepods, like many other crustaceans, have a fascinating life cycle. They start as eggs, which hatch into nauplius larvae. These larvae undergo several molting stages (usually 6), called naupliar stages, growing and developing with each molt. After passing through these stages, they enter the copepodite phase, which comprises five additional molts. Each stage brings them closer to their adult form, where they finally develop reproductive organs. As adults, they play a crucial role in the aquatic food web, serving as a vital food source for many larger marine organisms.

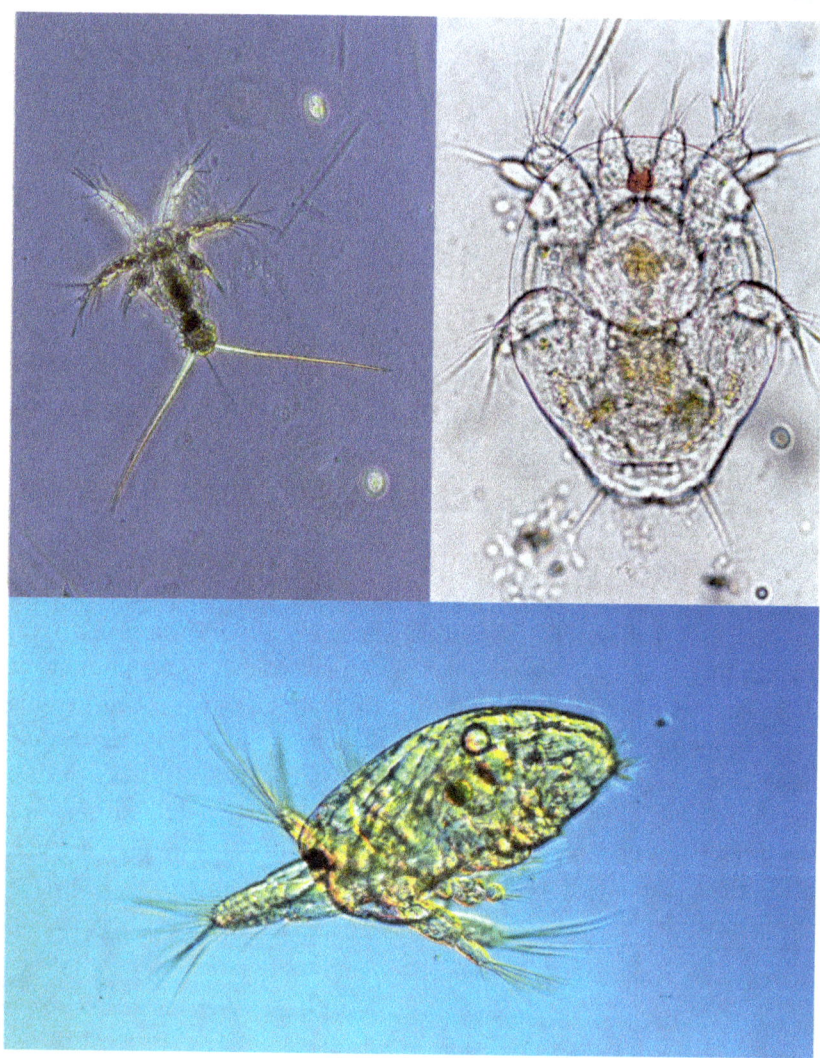

Copepod nauplii

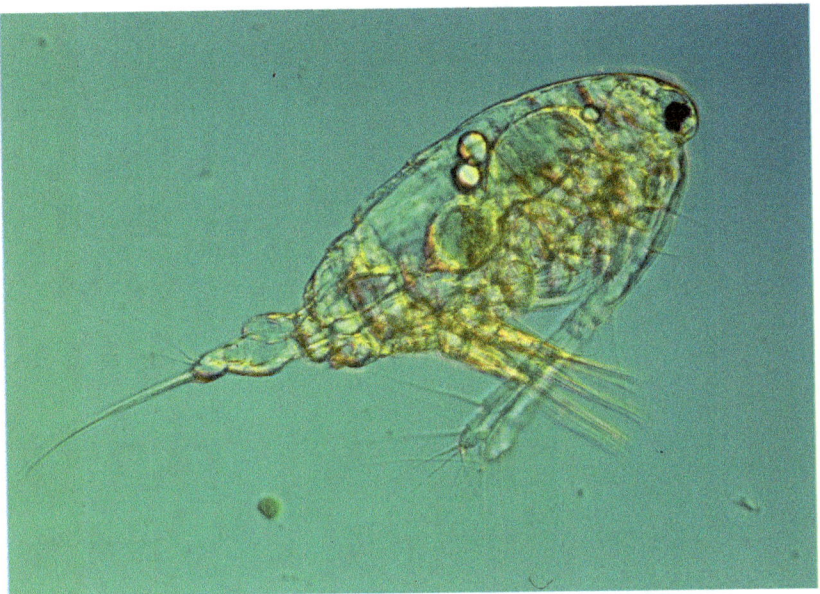

Copepodite

Cladocerans, also known as water fleas, are small, mostly freshwater crustaceans (although important in marine environments as well during summer), with *Daphnia* being one of the most well-known freshwater genera. They are filter feeders, consuming algae, bacteria, and detritus, and are crucial in freshwater ecosystems, forming the diet of fish larvae and other invertebrates. Cladocerans typically reproduce parthenogenetically (i.e., with no males). However, when environmental conditions deteriorate, males appear in the population, allowing for sexual reproduction. The eggs produced through male fertilization (ephippia) are resistant and adapted to endure the challenging conditions ahead.

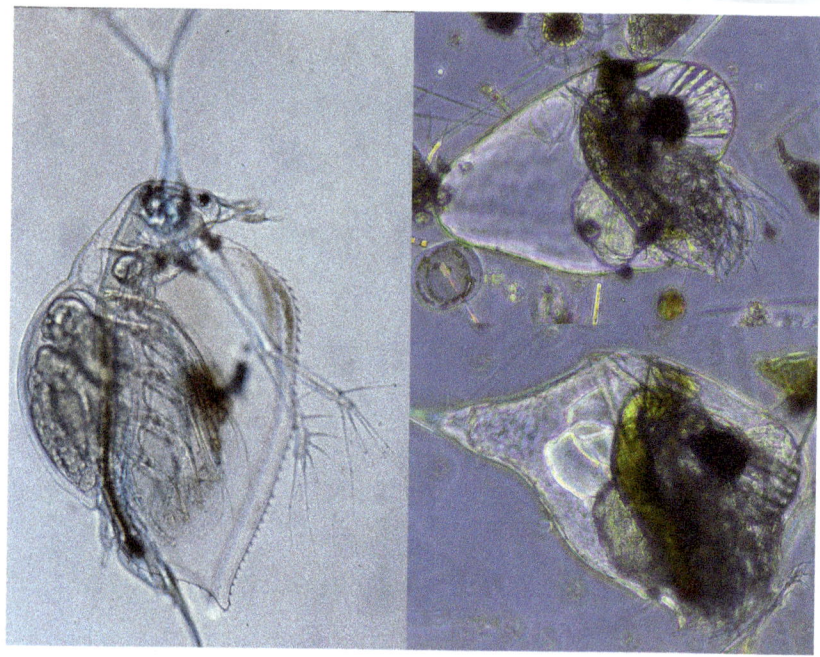

Marine cladocerans

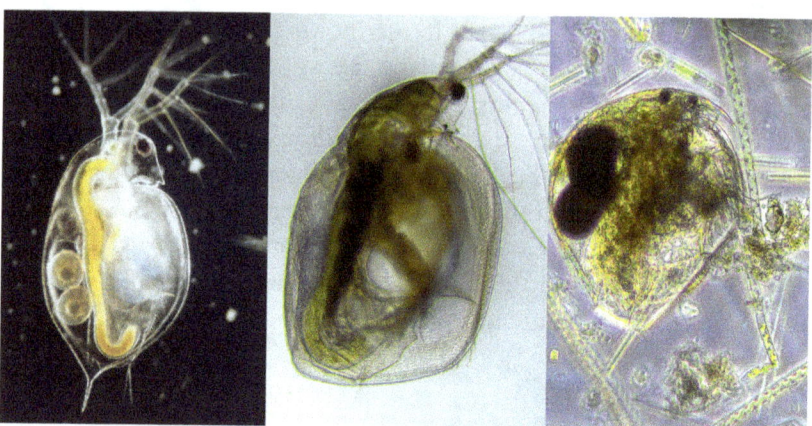

Freshwater cladocerans

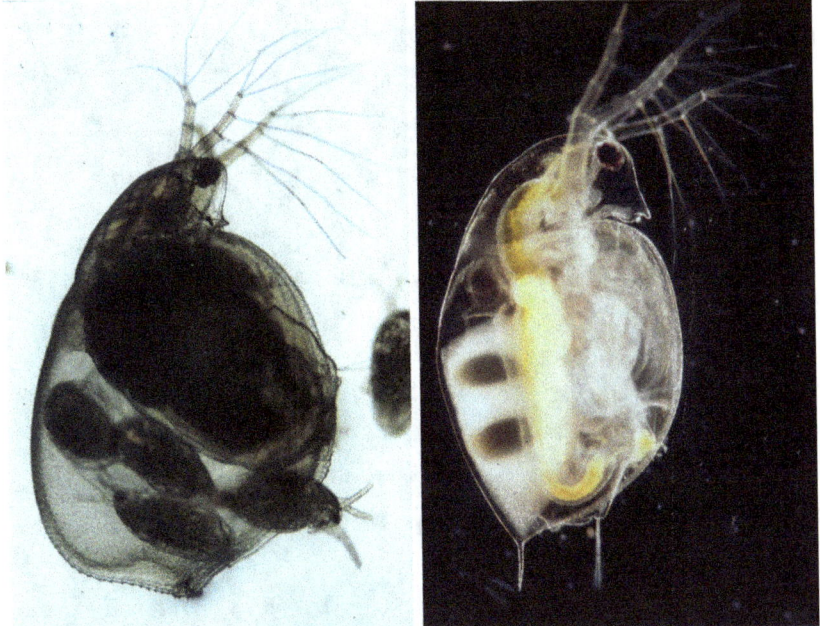

Parthenogenetic (left) and sexual (female with ephippia, right) reproduction

Euphausiids, commonly known as krill, are small, shrimp-like crustaceans found primarily in marine environments. Krill are essential in marine food webs, especially in polar regions, where they are the primary food source for whales, seals, penguins, and many fish species.

Meganyctiphanes norvegica
Hopcroft/UAF/CoML

5000 µm

Krill. Image R. Hopcroft

Ostracods are small crustaceans found in marine and freshwater environments. They have a bivalve-like shell that encloses their bodies, resembling tiny clams. Ostracods are diverse, with many species serving as plankton, detritivores, or parasites. Their fossilized remains are often used in paleontology to study ancient environments.

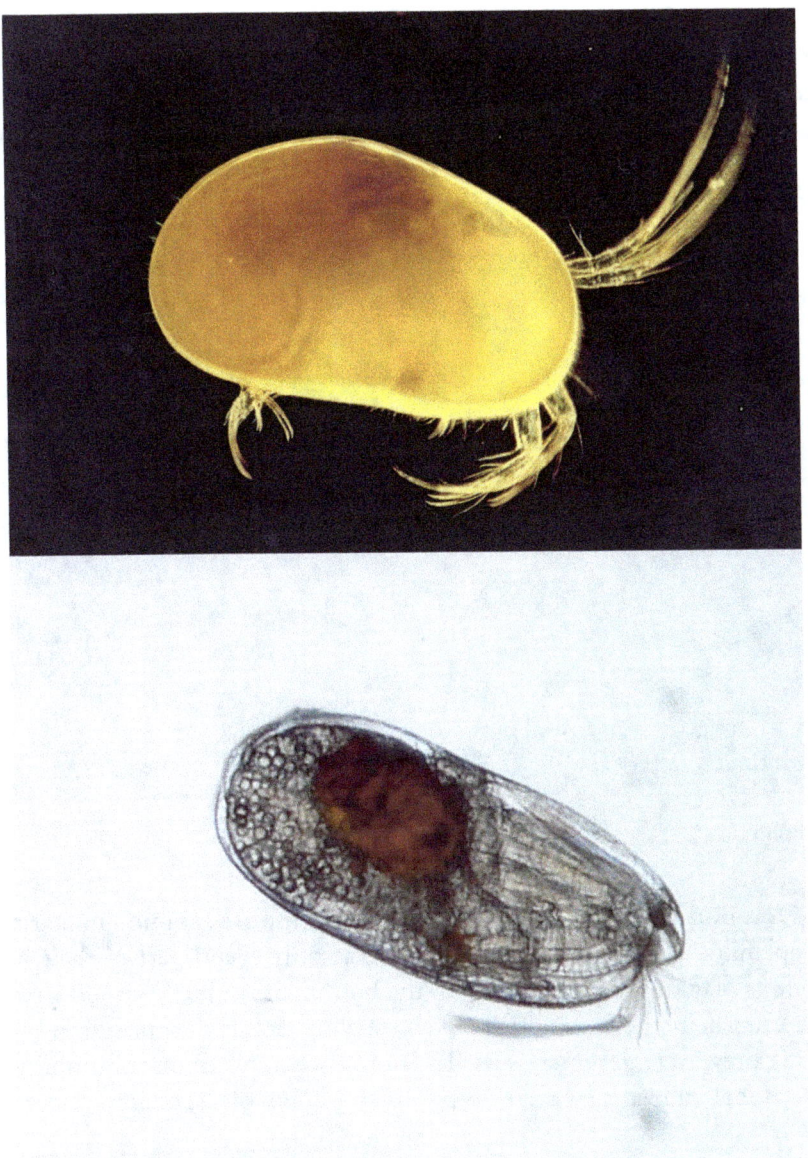

Ostracods (above, freshwater; below, marine)

Amphipods are small, laterally compressed crustaceans found in both marine and freshwater environments, with species adapted to a wide range of habitats, including open water and benthic zones. Amphipods are scavengers and detritivores, feeding on dead organic matter, but some are also predatory. They play a key role in the decomposition process and are a food source for fish and larger invertebrates.

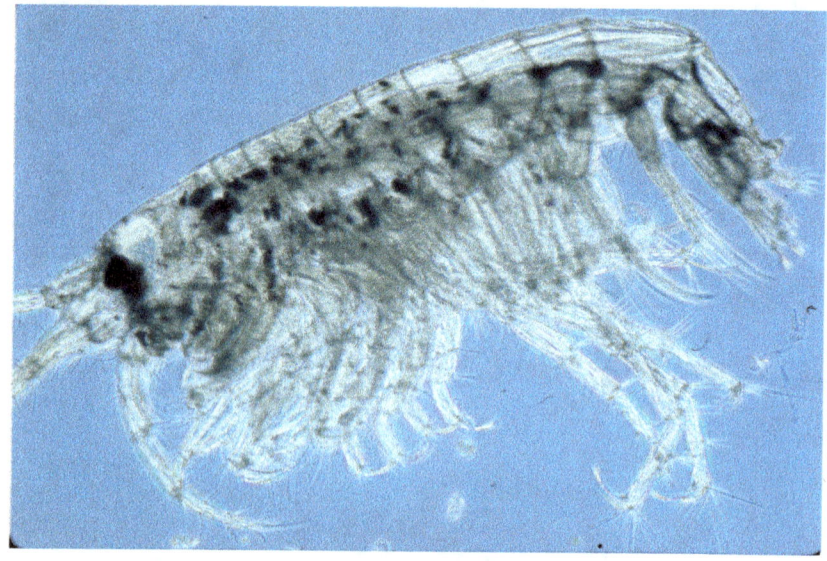

Amphipod

Decapod Larvae of crabs, lobsters, and shrimp are common in marine zooplankton. These meroplanktonic larvae only spend part of their life cycle as plankton before settling to the bottom as adults. Decapod larvae are crucial in the marine food web, feeding on smaller plankton and being prey for larger organisms. Their presence in plankton communities is seasonal and varies with the reproductive cycles of the adults.

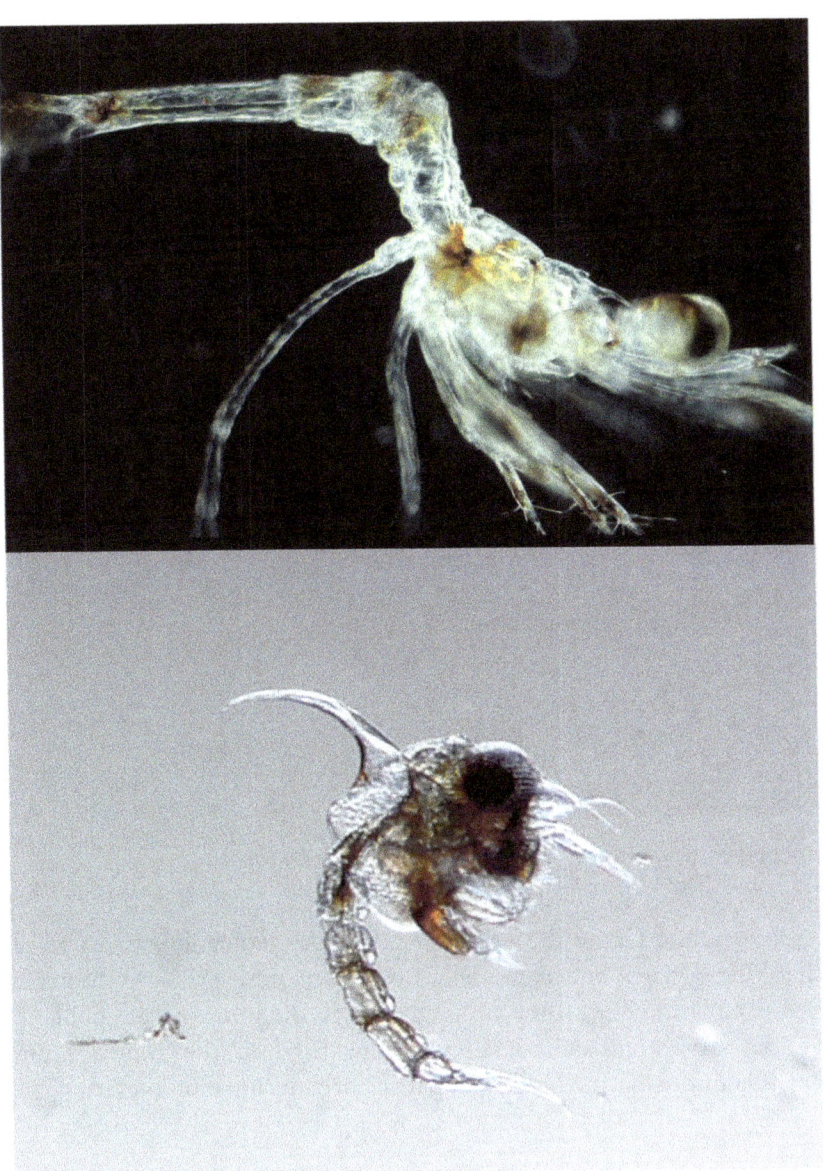

Decapod larva

Isopods are crustaceans that include both benthic and planktonic species, though their planktonic stage is typically limited to their larval forms. As meroplankton, isopod larvae contribute to the diversity of the plankton community and serve as prey for larger zooplankton and fish.

Copepod with a parasitic isopod attached to its body

Jellyfish and Ctenophores. These gelatinous zooplankton include jellyfish (Cnidarians) and comb jellies (Ctenophores), primarily found in marine environments. They are predators, feeding on smaller zooplankton, fish larvae, and sometimes even other jellyfish. They play a significant role in the marine food web, particularly during bloom events.

Ephyra of jellyfish

Adult jellyfish. Although they may be large, they are considered plankton

Ctenophore

Chaetognaths, or arrow worms, are slender, transparent predators found in both marine and freshwater environments. As active predators, arrow worms feed on copepods, small zooplankton, and fish larvae, playing a significant role as mid-level predators in aquatic ecosystems.

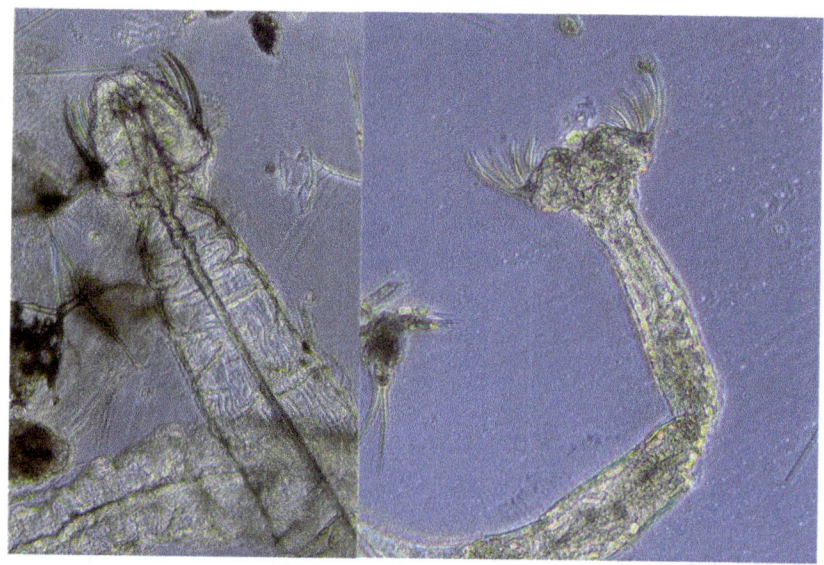

Chaetognaths

Larvaceans and pelagic tunicates: Larvaceans (Appendicularia) are small, tadpole-like tunicates that build a gelatinous house, while doliolums and salps are gelatinous tunicates, organisms that often form long chains. Both are predominantly marine. These organisms are filter feeders, capturing phytoplankton and small particles from the water. Salps can significantly impact carbon cycling by rapidly transporting carbon to deeper ocean layers.

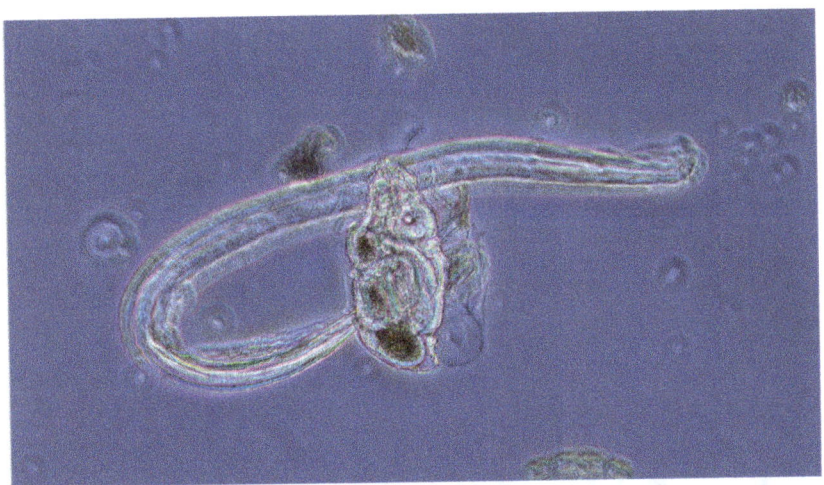

Appendicularia

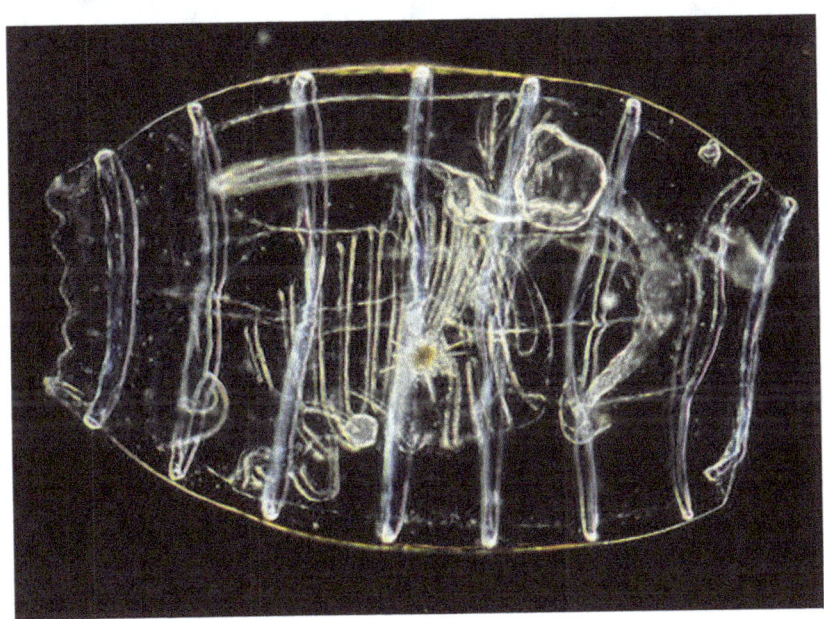

Doliolum

Planktonic mollusks are free-floating marine organisms, primarily composed of small gastropods like pteropods and heteropods. Unlike their benthic counterparts, these mollusks live in the water column and are often microscopic. They play a vital role in ocean ecosystems as both predators and prey, contributing to the marine food web and participating in carbon cycling by sequestering carbon when their calcium carbonate shells sink to the ocean floor. Some species, like shelled pteropods, are sensitive to ocean acidification, making them important indicators of climate change.

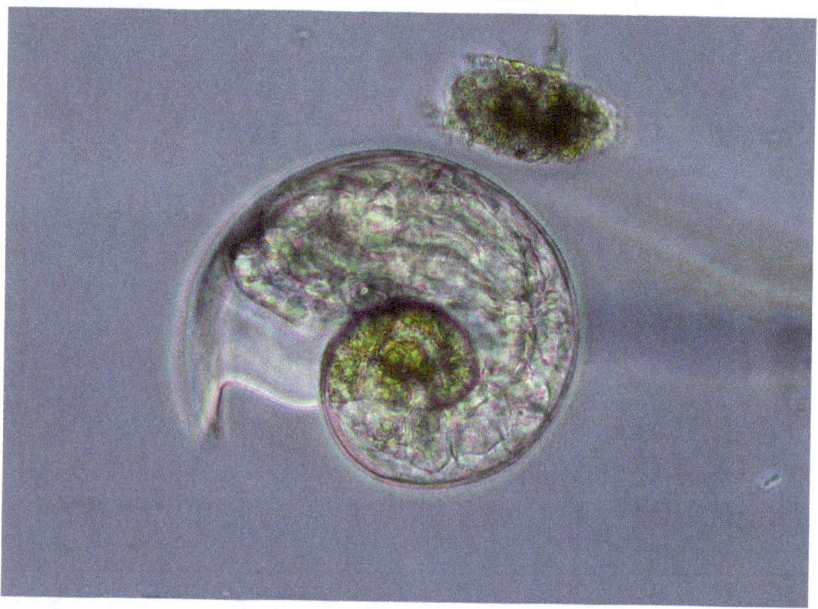

Planktonic snail

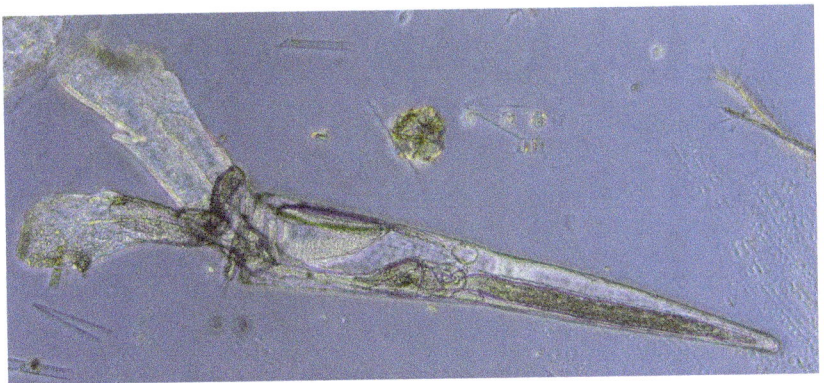

Pteropod, planktonic mollusk

Glossary[1]

Autotroph An organism capable of producing its own food through photosynthesis or chemosynthesis.

Biodiversity The variety of life in a particular habitat or ecosystem.

Biomass The total mass of organisms in a given area or volume.

Cilia Hair-like structures that aid in movement and feeding in some plankton species.

Chlorophyll A pigment found in plants and phytoplankton that is essential for photosynthesis.

Copepods Small crustaceans found in the plankton community, serving as a primary food source for many marine animals.

Dinoflagellates A group of single-celled phytoplankton characterized by two flagella for movement.

Diatoms A major group of microalgae with silica cell walls, known for their diverse and intricate shapes.

Eutrophication An excessive richness of nutrients in a body of water, often resulting in a dense growth of algae and depletion of oxygen.

---

[1] Understanding key terms and definitions is essential for studying plankton. This section provides a glossary of important terms you will encounter in your research.

Flagella Long, whip-like appendages used for movement in certain plankton species.

Heterotroph An organism that obtains its food by consuming other organisms.

Holoplankton Plankton that spend their entire life cycle as part of the plankton community.

Meroplankton Organisms that are planktonic only during their larval stage and later become benthic or nektonic.

Metazoan Organisms composed of multiple cells.

Micron or micrometer A unit of length equal to one-millionth of a meter, commonly used to measure plankton.

Mixoplankton An organism that shares both phytoplankton and zooplankton characteristics.

Phytoplankton Microscopic algae that live in aquatic environments and conduct photosynthesis.

Plankton Small and microscopic organisms that drift or float in the water, including phytoplankton and zooplankton.

Primary Production The creation of organic compounds from carbon dioxide through photosynthesis, primarily by phytoplankton in aquatic environments.

Protists A diverse group of eukaryotic microorganisms, including many plankton species.

Protozoan Unicellular heterotrophic microorganisms.

Silica A hard, glass-like compound found in the cell walls of diatoms.

Zooplankton The animal component of the plankton community, including small crustaceans and larvae of larger organisms. In many books (like this one) and scientific articles, the term zooplankton also includes protozoans.